T0213269

Science and Fiction

Science and Fiction – A Springer Series

This collection of entertaining and thought-provoking books will appeal equally to science buffs, scientists and science-fiction fans. It was born out of the recognition that scientific discovery and the creation of plausible fictional scenarios are often two sides of the same coin. Each relies on an understanding of the way the world works, coupled with the imaginative ability to invent new or alternative explanations—and even other worlds. Authored by practicing scientists as well as writers of hard science fiction, these books explore and exploit the borderlands between accepted science and its fictional counterpart. Uncovering mutual influences, promoting fruitful interaction, narrating and analyzing fictional scenarios, together they serve as a reaction vessel for inspired new ideas in science, technology, and beyond.

Whether fiction, fact, or forever undecidable: the Springer Series "Science and Fiction" intends to go where no one has gone before!

Its largely non-technical books take several different approaches. Journey with their authors as they

- Indulge in science speculation – describing intriguing, plausible yet unproven ideas;
- Exploit science fiction for educational purposes and as a means of promoting critical thinking;
- Explore the interplay of science and science fiction – throughout the history of the genre and looking ahead;
- Delve into related topics including, but not limited to: science as a creative process, the limits of science, interplay of literature and knowledge;
- Tell fictional short stories built around well-defined scientific ideas, with a supplement summarizing the science underlying the plot.

Readers can look forward to a broad range of topics, as intriguing as they are important. Here just a few by way of illustration:

- Time travel, superluminal travel, wormholes, teleportation
- Extraterrestrial intelligence and alien civilizations
- Artificial intelligence, planetary brains, the universe as a computer, simulated worlds
- Non-anthropocentric viewpoints
- Synthetic biology, genetic engineering, developing nanotechnologies
- Eco/infrastructure/meteorite-impact disaster scenarios
- Future scenarios, transhumanism, posthumanism, intelligence explosion
- Virtual worlds, cyberspace dramas
- Consciousness and mind manipulation

More information about this series at http://www.springer.com/series/11657

Andrew May

Fake Physics: Spoofs, Hoaxes and Fictitious Science

 Springer

Andrew May
Crewkerne, UK

ISSN 2197-1188   ISSN 2197-1196   (electronic)
Science and Fiction
ISBN 978-3-030-13313-9   ISBN 978-3-030-13314-6   (eBook)
https://doi.org/10.1007/978-3-030-13314-6

Library of Congress Control Number: 2019934804

Cover illustration: Scientific vector seamless pattern with math and physical formulas, chemistry plots and graphic schemes, shuffled together. Endless math texture.
By Marina Sun/shutterstock.com

This Springer imprint is published by the registered company Springer Nature Switzerland AG.
The registered company address is: Gewerbestrasse 11, 6330 Cham, Switzerland

Preface

In recent years, I've become fascinated by the overlaps—and occasionally fuzzy boundaries—between subjects that aren't normally mentioned in the same breath. That was the idea behind my previous contributions to Springer's "Science and Fiction" series: *Pseudoscience and Science Fiction* (2017) and *Rockets and Ray Guns: The Sci-Fi Science of the Cold War* (2018). The first, as the title suggests, looked at overlaps between science fiction (SF) and pseudoscientific writings on subjects like UFOs, antigravity, and telepathy, while the second considered the (even more surprising) overlaps between SF and real-world science during the Cold War period.

Soon after I finished writing *Rockets and Ray Guns*, the series editor Christian Caron drew my attention to the considerable number of spoof papers—mostly written as April Fool jokes—to be found in the arXiv online preprint repository. Superficially the papers look just like any others on arXiv, written in traditional academic style, and formatted as if they were scheduled for publication in a professional science journal. As with arXiv as a whole, the spoof papers tend to deal with cutting-edge physics and related fields—the difference being that the research reported is totally spurious and often very funny.

I was reminded of a number of other spoofs written in formal academic style. Best known in the SF community is Isaac Asimov's "The Endochronic Properties of Resublimated Thiotimoline", published in *Astounding* magazine in 1948. Then there's Alan Sokal's hoax paper "Transgressing the Boundaries: Towards a Transformative Hermeneutics of Quantum Gravity", which was published by a social sciences journal in 1996 without the editors realizing it was a spoof. More recently, there have been numerous well-publicized cases of so-called predatory journals—which charge authors for editing and reviewing services they never actually deliver—accepting spoof papers that consist of little more than gibberish.

Both Chris and I felt there was potential for another book here, but it took a while to settle on the best format for it. If we concentrated too much on spoofs originally written for an audience of professional academics, there was a danger that general readers wouldn't find them funny. On the other hand, the best of the academic spoofs—with the help of a judicious amount of background explanation—can be appreciated by anyone. The same, of course, is true of the kind of spoof "technobabble" often found in SF—which can be especially convincing when it's written by authors who are also professional scientists.

I realized this was turning into another "overlap" book like the first two— in this case, the overlap (and fuzzy boundary) between SF and the more whimsical fringes of real science. As well as the outright spoofs already mentioned, the latter includes highly speculative concepts like faster-than-light tachyons and the "multiverse", as well as the numerous "thought experiments" used to explain difficult ideas from relativity and quantum theory.

So the result was *Fake Physics*—a deliberately broad term that encompasses a range of different topics. The main criteria for inclusion are that the "fake physics" should be intentional (on the part of the authors) and entertaining— not just to professional physicists but to ordinary SF readers as well. Here is a quick rundown of the book's contents.

The first chapter, "Science Fiction Posing as Science Fact", starts on what should be familiar territory to many readers: Asimov's original Thiotimoline spoof, as well as a number of follow-ups to it written by Asimov and others. It also takes a broader look at various ways in which fiction writers try to persuade readers they're actually reading non-fiction.

The second chapter "The Relativity of Wrong" (a phrase coined by Asimov) takes a step back to look at the way real science works—and how aspects of its methodology, specifically the formulation of hypotheses, can easily be twisted to create science-fictional (or in some cases, science-factual) "fake physics".

The third chapter looks at "The Art of Technobabble". Professional physics has both a language of its own—a mixture of jargon and mathematics—and a literary style, namely that of the academic paper. The latter is particularly important, because both the promulgation of scientific ideas and the furtherance of scientific careers depend on scientists publishing their results.

This brings us to the next chapter: "Spoofs in Scientific Journals". Some of these are indistinguishable in style, format, and intent from Asimov's thiotimoline piece—the only difference being that they appeared in publications (either serious or not so serious) aimed at professional scientists rather than SF readers.

Not surprisingly, the appearance of spoofs has a peak around the 1st of April each year—to the extent that the "April Fool" phenomenon requires a whole chapter to itself. Some of these spoofs appeared in traditional print media, but—as already mentioned—their real home today is the online arXiv repository.

So far, all the spoofs discussed have been purely for fun. The ones in the next chapter—"Making a Point"—are funny too, but the authors had another reason for writing them besides making people laugh. This is where you'll find the Sokal hoax and various "sting operations" against unscrupulous predatory journals.

The final chapter is called "Thinking Outside the Box". While the "fake physics" here undoubtedly has fake aspects—and is often highly entertaining—it nevertheless carries a serious scientific message. There's scientific debunking of paranormal claims, thought experiments like the relativistic "twin paradox", and discussions of other universes governed by different physical laws.

There are some types of "fake physics" you won't find in this book, such as frauds perpetrated for financial or professional gain. That's an unsavoury subject, and it fails to meet our "must be entertaining" criterion. Another exclusion is pseudoscience—for the simple reason that it's already been covered in *Pseudoscience and Science Fiction*.[1]

One final caveat—the word "physics" has been taken in a broader-than-usual sense, to include all the physical sciences—and even some other sciences when we simply couldn't resist it. Thus, you'll find things like a spoof NASA report on sex experiments in space and a hoax paper on the clinical pathophysiology of *Star Wars* style "midichlorians". Surely no one can blame us for including those!

Crewkerne, UK   Andrew May

---

[1] Though I couldn't resist including a spoof of my own on this subject, "Science for Crackpots", which first appeared in *Mad Scientist Journal* and is included as a short appendix to this book.

Contents

Science Fiction Posing as Science Fact   1
    The Thiotimoline Saga   1
    Beyond Thiotimoline   9
    A Venerable Tradition   17
    References   21

The Relativity of Wrong   23
    How Science Works   23
    Science-Fictional Hypotheses   28
    Not Even Wrong?   39
    The Pauli Effect   44
    References   47

The Art of Technobabble   49
    The Lingo of the Sciences   49
    Damned Lies and Statistics   54
    Publish or Perish   57
    Automatic Paper Generators   62
    References   67

Spoofs in Science Journals   69
    Spoof Papers   69
    Spoof Journals   77
    The Ig Nobel Prize   81

A Touch of Humour   84
References   90

April Fool   93
A Day When Nobody Believes Anything   93
The ArXiv Spoofs   99
A Joke Too Far?   106
References   109

Making a Point   113
Scientific Writing, the Lazy Way   113
Predatory Journals   116
Science Wars   121
Fake News   128
References   134

Thinking Outside the Box   137
Science Fact Posing as Science Fiction?   137
Thought Experiments   142
Different Physics   147
The Spectrum of Fake Physics   157
References   159

Appendix: Science for Crackpots   163

Science Fiction Posing as Science Fact

Abstract In 1948, the magazine *Astounding Science Fiction* printed a piece by Isaac Asimov about a fictitious substance, thiotimoline, which dissolves a second or so before it's added to water. The remarkable thing about "The Endochronic Properties of Resublimated Thiotimoline" was that it took the form of a spoof research paper rather than a short story. It started a trend that was continued over the years by Asimov and others, including quite a few professional scientists. This chapter unravels the story of thiotimoline and its successors—and traces the origins of such spoofs to earlier efforts to pass fiction off as fact, such as the use of spurious but real-looking maps in *Gulliver's Travels*.

The Thiotimoline Saga

In 1948 Isaac Asimov was 28 years old and already one of the world's leading writers of science fiction (SF). Over 40 of his stories had appeared in the various SF magazines of the time, including the most prestigious of all of them, *Astounding Science Fiction*, edited by John W. Campbell. Asimov's contributions to *Astounding* included most of the material that was later collected in the *Foundation* trilogy and the book *I, Robot*.

At the same time Asimov was coming to the end of a three year postgraduate course in chemistry at Columbia University in New York. As well as all that fiction, he was busy writing a thesis called "The kinetics of the reaction inactivation of tyrosinase during its catalysis of the aerobic oxidation of catechol". It just so happens that catechol is a compound that dissolves very readily in water—the instant it hits the surface—and this fact fascinated Asimov. He later recounted in his book *The Early Asimov*:

© Springer Nature Switzerland AG 2019
A. May, *Fake Physics: Spoofs, Hoaxes and Fictitious Science*, Science and Fiction,
https://doi.org/10.1007/978-3-030-13314-6_1

Idly, it occurred to me that if the catechol were any more soluble than it was, it would dissolve before it struck the water surface. Naturally, I thought at once that this notion might be the basis for an amusing story. It occurred to me, however, that instead of writing an actual story based on the idea, I might write up a fake research paper on the subject and get a little practice in turgid writing [1].

The result was a spoof research paper, "The Endochronic Properties of Resublimated Thiotimoline", written in the meticulous, impersonal style of the scientific journals of the time (or of today, for that matter). In spite of that, Asimov submitted the piece to his favourite SF magazine, John Campbell's *Astounding*. Fortunately Campbell—who had trained as a scientist himself—loved the joke, and printed Asimov's spoof in the March 1948 issue.

To be honest, it's really quite a thin joke. If it had been written up in the form of an ordinary short story, without the addition of other factors, it would have been a weak and forgettable one. The idea is simply that a fictitious substance called "thiotimoline" dissolves in water a second or so before it's actually added. What makes the six-page piece so memorable—and genuinely very funny—is its ostensibly serious format, complete with numerical tables, diagrams and a formal list of references at the end. All these things are absolutely standard in scientific papers, but they'd never been seen before in a work of fiction.

"The Endochronic Properties of Resublimated Thiotimoline" had a huge impact when it appeared. Quoting from *The Early Asimov* again:

Although "Thiotimoline" appeared in *Astounding*, as did all my stories of the time, it received circulation far outside the ordinary science fiction world. It passed from chemist to chemist, by way of the magazine itself, or by reprints in small trade journals, or by copies pirated and mimeographed, even by word of mouth. People who had never heard of me at all as a science fiction writer, heard of thiotimoline. It was the very first time my fame transcended the field.

The thiotimoline piece highlighted a strange-but-true fact about spoofs in general: no matter how outrageous they are, if they're written in a superficially factual style, some people will take them for the truth. Asimov goes on:

I was told that in the weeks after its appearance the librarians at the New York public library were driven out of their minds by hordes of eager youngsters who demanded to see copies of the fake journals I had used as pseudo references [2].

The author of this book was lucky enough to meet Isaac Asimov in person soon after *The Early Asimov* came out in paperback (see Fig. 1).

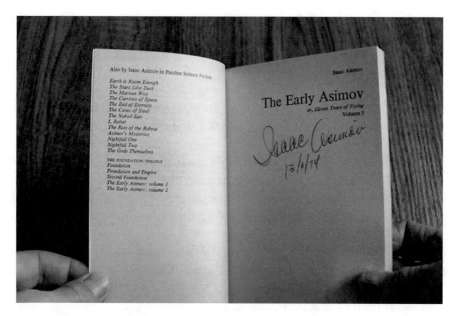

Fig. 1 The author's copy of *The Early Asimov, volume 3*, signed by Isaac Asimov in 1974

Here is opening of "The Endochronic Properties of Resublimated Thiotimoline", which gives a good flavour of its deliberately turgid style:

The correlation of the structure of organic molecules with their various properties, physical and chemical, has in recent years afforded much insight into the mechanism of organic reactions, notably in the theories of resonance and mesomerism as developed in the last decade. The solubilities of organic compounds in various solvents has become of particular interest in this connection through the recent discovery of the endochronic nature of thiotimoline.[1]

It has been long known that the solubility of organic compounds in polar solvents such as water is enhanced by the presence upon the hydrocarbon nucleus of hydrophilic—i.e. water-loving—groups, such as the hydroxy (–OH), amino (–NH$_2$), or sulphonic acid (SO$_3$H) groups. Where the physical characteristics of two given compounds—particularly the degree of subdivision of the material—are equal, then the time of solution—expressed in seconds per gram of material per millilitre of solvent—decreases with the number of hydrophilic groups present. Catechol, for instance, with two hydroxy groups on the benzene nucleus dissolves considerably more quickly than does phenol with only one hydroxy group on the nucleus. Feinschreiber and Hravlek[2] in their studies on the problem have contended that with increasing hydrophilism, the time of solution approaches zero.

That this analysis is not entirely correct was shown when it was discovered that the compound thiotimoline will dissolve in water—in the proportions of 1 g/mL—in minus 1.12 seconds. That is, it will dissolve before the water is added [3].

The superscripts 1 and 2 in the above excerpt refer to the first two fictitious references (of nine in total) listed at the end of Asimov's article:

1. P. Krum and L. Eshkin, *Journal of Chemical Solubilities*, *27*, 109–114 (1944), "Concerning the Anomalous Solubility of Thiotimoline"
2. E. J. Feinschreiber and Y. Hravlek, *Journal of Chemical Solubilities*, *22*, 57–68 (1939), "Solubility Speeds and Hydrophilic Groupings"

As well as academic-style references, Asimov's thiotimoline paper incudes equally academic-looking tables and diagrams. Examples of these are shown Figs. 2 and 3 respectively.

In broad terms, there are two approaches to writing science fiction. In the commonest approach the author wants to tell a particular story, or make a particular point, and invents whatever fictional science is necessary in order to do that. Insofar as "The Endochronic Properties of Resublimated Thiotimoline" is SF, it falls in this category.

The second type of SF is more like real science in the way it works. In this case the author starts by making up a fictitious piece of science, and then thinks through all the possible consequences of it. Occasionally a piece of fictitious science that was originally created in the first way makes such an impact on the SF community that it's subsequently developed in the second way. That's what happened in the case of thiotimoline. It developed a life of its own, the later course of which was gradually worked out—by Asimov and others—over a period of many years.

The fact is that, if a substance like thiotimoline really existed, it would have a number of important practical applications. Asimov drew attention to one

Purification Stage	Average "T" (12 observations)	"T" extremes	% error
As Isolated	−0.72	−0.25; −1.01	34.1
First recrystallization	−0.95	−0.84; −1.09	9.8
Second recrystallization	−1.05	−0.99; −1.10	4.0
Third recrystallization	−1.11	−1.08; −1.13	1.8
Fourth recrystallization	−1.12	−1.10; −1.13	1.7
First resublimation	−1.12	−1.11; −1.13	0.9
Second resublimation	−1.122	−1.12; −1.13	0.7

Fig. 2 One of the tables from Asimov's original "Thiotimoline" paper (source: Internet Archive)

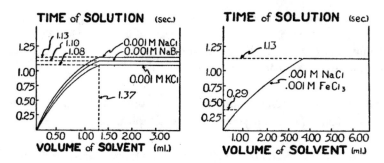

Fig. 3 Two of the diagrams from the original "Thiotimoline" paper (source: Internet Archive)

of these (though not necessarily the most important) in a follow-up piece in the same style called "The Micropsychiatric Applications of Thiotimoline". This appeared in the December 1953 issue of *Astounding Science Fiction*, with the opening line: "Some years ago, the unusual endochronic properties of purified thiotimoline were first reported in this journal". There's then a jokey endnote reference to:

Asimov, I. "The Endochronic Properties of Resublimated Thiotimoline", *Journal of Astounding Science Fiction*, 50 (#1), 120–125, (1948)

This second thiotimoline "paper" focuses on a potential paradox: what if a researcher is undecided as to whether to add thiotimoline to water or not? Will it still dissolve in advance? According to Asimov, the result depends on the researcher's willpower:

With ample supplies of thiotimoline of extreme purity finally made available by the use of endochronic filtration, it became possible to determine the effect of human will upon the negative time of solution—i.e. the endochronic interval—and, conversely, to measure the strength of the human will by means of thiotimoline...

It was early observed, for instance, that strong-willed, incisive personalities achieved the full endochronic interval when adding water by hand. Having made up their minds, in other words, that they were going to add the water, no doubts assailed them and the final addition was as certain as though it had been mechanically arranged. Other individuals, of a more or less hesitating, self-deprecatory nature, yielded quite different results. Even when expressing themselves as entirely determined to add the water in response to a given signal, and though assuring us afterward that they had felt no hesitation, the time of negative solution decreased markedly. Undoubtedly, their inner hesitation was so deeply bound with their unconscious mind and with super-ego-censored infantile traumas that they were completely unaware of it in any conscious manner. The importance of such physical demonstrations, amenable to quantitative treatment, to the psychiatrist is obvious.

Asimov goes on to describe an unexpected bonus to all this, in the form of a useful application to the psychology of schizophrenia:

In the case of one subject, however, J. G. B., it was found that, strangely enough, there was a perceptible time during which part of the thiotimoline had dissolved and part had not. … The subject, however, when subjected to thoroughgoing psychoanalysis, promptly displayed hitherto undetected schizophrenic tendencies. The effect on the endochronic interval of two personalities of differing degrees of self-confidence within a single mind is obvious [4].

"The Micropsychiatric Applications of Thiotimoline" is also notable for providing a quasi-scientific explanation for thiotimoline's peculiar properties:

In the 19th century, it was pointed out that the four valence bonds of carbon were not distributed toward the points of a square … but toward the four vertices of a tetrahedron. The difference is that in the first case, all four bonds are distributed in a single plane, while in the second, the bonds are divided, two and two, among two mutually perpendicular planes. … Now once more we can broaden our scope. We can pass from the "tetrahedral carbon atom" to the "endochronic carbon atom" in which the two planes of carbon valence bonds are not both spatial in the ordinary sense. One, instead, is temporal. It extends in time, that is. One bond extends toward yesterday and one toward tomorrow.

As a consequence of this, "a small portion of the thiotimoline molecule exists in the past and another small portion in the future" [4].

Actually, Asimov's second thiotimoline paper marked the third appearance in *Astounding* magazine of the fictitious substance he'd invented. It had previously cropped up in a spoof article by another author in the September 1949 issue. This took the form of a ten-page "Progress Report" by John H. Pomeroy—not an SF writer, but a professional scientist who happened to be a fan of *Astounding*.

Pomeroy's spoof is cast in the form of a progress report for the third quarter of 1949, from the fictitious "Northeastern Divisional Laboratories" to the National Council on Science and Technology in Washington DC. The piece contains a number of satirical items, most of which come across as very weak jokes today. However, near the beginning Pomeroy writes:

Work on the determination of the structure, the synthesis, and further applications of thiotimoline has been carried on rapidly under the stimulus of a rapidly expanding staff. Scientific interest in this material has remained high ever since the preliminary announcements of its unique endochronic properties by Dr Asimov; we are fortunate in having his services as Acting Thiotimoline Co-ordinator.

Pomeroy goes on to talk about "selenotimoline, the selenium homolog of thiotimoline". In some ways, this is even more interesting than its predecessor:

Not only does this material possess the endochronic properties of thiotimoline but shows as well a selective reactivity to light that is not too surprising considering the known sensitivity of selenium itself. Selenotimoline darkens on exposure to light before the photons strike it, possibly by some amplification of the preceding probability wave function. The Polaroid Corporation has shown a great deal of interest in this application, and at present is working on a modification of the Land 60-second camera which will give the photographer a positive print of a scene before he snaps the shutter. The potential value of this invention in saving film that might have been taken of undesired subjects is, of course, obvious. Part of this work, however, is at present under military secrecy regulations because of the interest of the Air Force in applying these phenomena to directors and predictors for anti-aircraft fire [5].

Another contribution to the thiotimoline saga came from married British scientists Anne McLaren and Donald Michie. They're both important enough to have their own articles on Wikipedia, the former being "a leading figure in developmental biology" [6] and the latter an expert in artificial intelligence who worked at Bletchley Park with Alan Turing during World War Two [7].

In 1959, McLaren and Michie produced a spoof paper, "New Experiments with Thiotimoline", which was published in the *Journal of Irreproducible Results* (JIR). More will be said about this august periodical in a later chapter ("Spoofs in Science Journals")—but suffice to say that it was basically a spoof in itself, with much of the contents given over to what its founding editor, Alexander Kohn, described as "half-baked scientific ideas … carried as far as possible to their practical or logical conclusions". Expanding on this, he went on:

As an example, we may cite the papers of Asimov and later of McLaren and Michie (JIR, vol. 8; 1959, p. 27) on the discovery, the properties and the uses of thiotimoline. Thiotimoline is a substance which dissolves just before water is added to it. This peculiar property is due to thiotimoline's having in its structure one carbon atom sticking out into the fourth dimension. Thiotimoline found important applications for the prediction of weather: if thiotimoline in a reaction vessel dissolves one second before the addition of water, then a battery of 86,400 such vessels (60 × 60 × 24), linked so that each successively activates the next, would enable the exact, and perfect, prediction of rain yesterday [8].

This is a clever idea—and an example of the second kind of science-fictional thinking, in which a fictitious concept is thought through to all its logical consequences. In the same way that a number of low-voltage cells can be linked together to produce a high-voltage battery, so a "thiotimoline battery" can be constructed which gives a much longer anticipatory interval than the second or so allowed by a single sample of thiotimoline. It is this idea—and its application to weather forecasting—that was thought up by McLaren and Michie.

Before long, however, their work was cited by Asimov himself in his next thiotimoline article. In 1960, John Campbell changed the name of his magazine from *Astounding* to *Analog Science Fact & Fiction*, and Asimov's new piece appeared in the October issue that year. Called "Thiotimoline and the Space Age", it was slightly different in style from its predecessors. Rather than a formal scientific paper, it was presented in the form of a "transcript of a speech delivered at the 12th annual meeting of the American Chronochemical Society". In it, Asimov gives due credit both to the McLaren & Michie paper and to Alexander's Kohn's journal:

> Thiotimoline research graduated from what we might now call the "classical" stage to the "modern" with the development of the "telechronic battery" by Anne McLaren and Donald Michie of the University of Edinburgh. ... The original paper appeared only in the small, though highly respected, *Journal of Irreproducible Results*, edited by that able gentleman Alexander Kohn.

Asimov takes the British scientists' idea that a "device of not more than a cubic foot in volume can afford a 24-hour endochronic interval" and applies it, not to the prosaic subject of weather forecasting, but to one of the trendiest engineering problems of the time: predicting if a satellite launch will be successful or not.

> Suppose that four hours after launching, an automatic device on board the satellite telemeters a signal to the launching base. Suppose, next, that this radio signal is designed to activate the first element of a telechronic battery. Do you see the consequences? The sending of the signal four hours after launching can only

mean that the satellite is safely in orbit. If it were not, it would have plunged to destruction before the four hours had elapsed. If then, the final element of the telechronic battery dissolves today, we can be certain that there will be a successful launching tomorrow and all may proceed [9].

Asimov returned to the subject of thiotimoline one further time, in a story he wrote as tribute to John Campbell after the latter's death in 1971. Called "Thiotimoline to the Stars", this time it really was a straightforward SF story. It's set in the far future, after it has become possible to make large objects like an entire spaceship endochronic—a useful trick, because if done correctly it can precisely counterbalance the effects of relativistic time dilation (within the logic of Asimov's story it can, anyway).

For present purposes, the most interest passage occurs near the start of the story, when thiotimoline is described as having been "first mentioned in 1948, according to legend, by Azimuth or, possibly, Asymptote, who may, very likely, never have existed" [10]. As well as the sly reference to himself, Asimov works in a couple of in-jokes for the benefit of the scientists and engineers in his audience. Both "azimuth" and "asymptote" are real words, the former referring to an angular measurement in the horizontal plane, and the latter to a line which is tangent to a curve at infinity.

In more recent years, thiotimoline has been picked up by several other authors, sometimes in contexts not even dreamed of in 1948. For example, in 2002 the usually serious journal *Design and Test of Computers*, from the Institute of Electrical and Electronics Engineers, carried an article by Rick Nelson called "Yet Another Thiotimoline Application". In it, the author describes "a keyboard-to-computer interface that causes the computer to record my keystrokes 1.12 seconds before I type them". He goes on:

A thiotimoline keyboard's benefits to humanity would be vast. Such a keyboard would eliminate the dreaded disease of writer's block. … All that a writer with a thiotimoline keyboard needs to do is quickly copy down the words that appear on the computer screen 1.12 seconds before being typed [11].

Beyond Thiotimoline

One of the reasons Asimov's original 1948 article made such an impact is the fact that it appeared in John W. Campbell's *Astounding*. It was by far the most upmarket SF magazine of the 1940s (see Fig. 4), boasting a readership that included a large number of professional scientists and engineers. As SF writer

Fig. 4 Three science fiction magazines from 1947, with *Astounding* clearly displaying a more serious attitude than the other two (public domain images)

and critic Brian Aldiss put it: "the typical *Astounding Science Fiction* story was rather cold and impersonal in tone, and sometimes degenerated into a kind of illustrated lecture". He went on to add that "there were times when *Astounding* smelt so much of the research lab that it should have been printed on filter paper" [12].

Every issue of Campbell's magazine contained at least one factual article about science. When "The Endochronic Properties of Resublimated Thiotimoline" appeared in March 1948, the same issue also contained a perfectly serious article about the design of pressure suits. The two previous months had offered articles on servomechanisms and magnetrons, both representing state-of-the-art engineering at the time. The next two issues saw pieces on Mira-type variable stars and the "Electrical Robot Brain". Despite its sci-fi-sounding title, the latter dealt with a real piece of military hardware "which automatically controlled the fire of a battery of four 90 mm guns" [13].

By far the most controversial "factual" article that *Astounding* ever printed was on the subject of Dianetics, the precursor of scientology, by author L. Ron Hubbard. As notorious as the subject later became, its first exposition aimed at a wide audience appeared in the May 1950 issue of *Astounding*. Anticipating the scepticism of his readers, Campbell asked a medical doctor, Joseph A. Winter, to write a short introduction to ensure that, in Winter's words, "readers would not confuse Dianetics with Thiotimoline or with any other bit of scientific spoofing" [14].

Its non-fiction articles notwithstanding, the bulk of *Astounding Science Fiction* consisted, as the title suggests, of science fiction. That's a broad term, however. As well as stories set in the far future, or in outer space, and featuring a cast of obviously fictional characters, there's another type of SF that borders more closely on spoofs. These could perhaps be described as "tall tales"—set in a recognisably ordinary present-day, and narrated in the first person, but with one very far-fetched element.

Asimov produced a good example of this sub-genre early in his career, in the form of "Super-Neutron"—originally published in another magazine, *Astonishing Stories*, in September 1941. It ostensibly describes a real-world occurrence at the "17th meeting of the honourable society of Ananias"—a social group supposedly devoted to the telling of tall stories:

> It was quite a complicated process, with strict parliamentary rules. One member spun a yarn each meeting as his turn came up, and two conditions had to be adhered to. His story had to be an outrageous, complicated, fantastic lie; and, it had to sound like the truth.

The particular story in question concerns the discovery by an astronomer of a planet-sized "super-neutron"—a wild idea which brought together then-topical aspects of both astrophysics and subatomic physics. It ends with the following observation by the narrator:

> I think he should have been disqualified after all. His story fulfilled the second condition; it sounded like the truth. But I don't think it fulfilled the first condition. I think it was the truth! [15]

The following year, another of Asimov's tall stories, "Time Pussy", appeared in the April issue of *Astounding* under the pen-name of George E. Dale. It was one of a number of such tales the magazine printed under the heading of "Probability Zero" (see Fig. 5), as Asimov explained in *The Early Asimov*:

> Campbell told me his plan for starting a new department in *Astounding*, one to be called "Probability Zero". This was to be a department of short-shorts, 500–1000 words, which were to be in the nature of plausible and entertaining lies [16].

In actuality, most of the "Probability Zero" pieces were fairly conventional short stories that couldn't be mistaken for anything but fiction. A few, however, took the form of short, thiotimoline-style spoofs. One of the best examples is "The Image of Annihilation" by Jack Speer, which appeared in the

Fig. 5 *Astounding Science Fiction*'s "Probability Zero" feature consisted of scientific-sounding tall tales—"lies" that might just be the truth (source: Internet Archive)

August 1942 issue of *Astounding*. Its narrator tells of a physics-based invention that sounds quite plausible in its way:

> I am privileged to supply final confirmation of the truth of the wave theory—you know: the idea that electrons are just etheric vibrations, like eddies in a stream of water. The results I secured by working on this hypothesis admit of no other conclusion.
>
> It is well known that in the case of sound—which is a vibration of the air, similar to vibrations of the ether like light and radio—when two sound waves of certain pitch are superimposed on each other so that the crests of the one fall exactly on the troughs of the other, they will automatically cancel out into silence. It occurred to me that I might be able to do the same thing with etheric vibrations, producing in a restricted area an utter absence of light, radio, and even matter.
>
> It was a simple thing to give a metal mirror a special coating to slightly change its reflecting properties so that it would reflect back the vibrations of material electrons, and a mechanical device on the back of it permitted bending the mirror out of focus except when I wanted to use it. Now, the reflections of the electrons directly in front of the mirror were exact duplicates of the original electrons, except that they came back upon them in reverse order. Naturally, the two waves cancelled each other out.

The author goes on to describe various uses he made of his matter-cancelling invention—including "trimming hedges and disposing of old razor blades and bill collectors" [17].

Campbell also printed a few longer spoofs in a similar vein—often designed to have a particular appeal to the numerous scientists and engineers among his readers. A good example, from the 1961 issue of *Analog*, is "An Introduction to the Calculus of Desk-Clearing" by Maurice Price. This describes the way desk clutter tends to increase exponentially with time t after any attempt at desk-clearing. It includes an equation which might look baffling to ordinary readers, but would be instantly understandable to most scientists and engineers:

$$C = K_1 \exp(K_2 t)$$

As Price explains:

In this equation K_1 is the constant of confusion and K_2 is the coefficient of chaos. These may vary from desk to desk and from engineer to engineer, but the general form of the curve is not altered. Note that the amount of work to be done does not affect the curve at all [18].

Analog's interest in spoofs didn't end with Campbell's death in 1971. In fact a later editor printed a special "spoof" issue, dated Mid-December 1984. From a physics perspective, the most interesting item in this issue is a jokey thiotimoline-style paper credited to A. Held, P. Yodzis & E. Zechbruder and titled "On the Einstein-Murphy Interaction". Like any real academic paper it has an abstract, which in this case is short and to the point:

This paper is a first attempt to reconcile the two great concepts of 20th century physics: Einstein's theory of general relativity, and Murphy's law.

Murphy's law, of course, enshrines the folk wisdom that "anything that can go wrong will go wrong". One of the most commonly cited examples is the notion that a slice of bread will always fall butter-side down. This is the scenario addressed by the authors, who adopt a mathematical model of the kind physicists habitually use to analyse such situations:

We begin by considering a loaf of bread which, for our purposes, will be considered to be a compact manifold admitting a well-behaved foliation. Each folium may be thickened and approximated by a rectangular parallelepiped of homogeneous density. Each folium (hereafter referred to as "slice") can be represented in the limit $\varepsilon \to 0$. ... With these reasonable assumptions, we find the centre of gravity of the slice to lie at its geometric centre:

$$d = \frac{\rho_b b^2 + \rho_j \left[(b+j)^2 - b^2 \right]}{2 (\rho_b b + \rho_j j)}$$

Having defined the static parameters of the problem, the authors move on to kinematics:

At time δ, the slice is (inadvertently) brushed by the hand and moves along the table with constant velocity v_0 in a direction perpendicular to the table edge so that side A remains parallel to the aforementioned edge. To obtain a reasonable upper limit for the value of v_0, measurements were carried out by B. Wälti of the Physics Department of the University of Bern. It was found that the maximum velocity attainable by the human hand when propelled by and remaining attached to its natural owner is of the order of 1500 cm/sec.

They proceed to derive suitable (and in fact perfectly sensible) equations of motion for the problem, and after presenting a number of solutions eventually conclude that "our results agree with Murphy's law in all cases provided that the slice-floor coefficient of friction is in excess of 1.65" [19].

Other SF magazines followed in *Astounding/Analog*'s footsteps. For example, *Galaxy* magazine often ran what they called "non-fact articles". A particularly physics-related one, by two of the leading SF writers of the 1960s—Thomas M. Disch and John Sladek—appeared in the February 1967 issue. Called "The Discovery of the Nullitron", it once again took the form of a spoof scientific report:

Whilst attempting a verification of Drake's classical "Massless Muon" experiment (the experiment in which a massless muon was annihilated, producing, as Hawakaja had earlier observed, the supposed "isotron"), a new particle was observed, having a mass of 0, a charge of 0 and a spin of 0. This particle has been termed the "nullitron"...

Though having no mass, the particle cannot be truly termed subatomic, for it appears to be about one metre in diameter, perfectly round and rather shiny. Its red colour can be explained by the well known "red-shift" or Doppler effect, caused by the fact that no matter from what vantage the particle is viewed, it seems to be retreating from the observer uniformly at the speed of light [20].

More than a decade later, John Sladek produced an ambitious spoof in the form of an entire book. By this time he had developed a sideline as a debunker of pseudoscience—anything from astrology to flying saucers—and was particularly annoyed by the way its purveyors twisted the methods and language of science to persuade readers that any old nonsense was true. To demonstrate how easy this was, Sladek emulated the practice himself in a book called *Judgement of Jupiter*, published in 1981 under the pseudonym of Richard A. Tilms. As the present author wrote in an earlier book in this series, *Pseudoscience and Science Fiction*:

> Although presented in the form of non-fiction, the book is clearly intended to be a tongue-in-cheek satire, and is as much a product of Sladek's imagination as any of his science fiction novels [21].

Most of the basic facts and data that Sladek presents in *Judgement of Jupiter* are genuine, with the "spoof" aspects lying in the far-fetched conclusions he draws from them. Here is an example, where he attempts to persuade the reader that terrestrial disasters are caused by planetary alignments:

> How can alignments of other planets cause earthquakes? That they do cause earthquakes is strongly suggested by coincidences like these:

> 26 August 1883. Mount Krakatoa's eruption east of Java, in the Dutch East Indies, the greatest volcanic disaster of the past 150 years, caused the deaths of some 36,000 persons and untold damage. ... The mighty eruption came less than a month after an extremely rare conjunction of Saturn with Pluto. The next conjunction of Saturn and Pluto comes in 1982.

> 18 April 1906. The famous San Francisco earthquake levelled four square miles of the city, killed over 500 and caused some 300 million dollars' damage. ... The catastrophe followed shortly after a conjunction of Jupiter with Pluto. The next conjunction of Jupiter and Pluto comes in 1981 [22].

Notice that Sladek gives himself away in the first paragraph, when he uses the cautious wording "is strongly suggested". A real pseudoscientist, of course, would have said "is proven beyond question".

Later in the book, Sladek makes use of another trick that can be played with facts and figures: taking any two sets of statistical data, scouring them for some kind of correlation, and then jumping to the conclusion that there is a cause-and-effect relation between them. Here is Sladek's example:

A link between the Full Moon and criminal psychosis can also be demonstrated in Britain, as [the following figure] shows. It compares the number of "abnormal" murders for England and Wales with the number of days of full moonlight for the years 1957–1967. "Abnormal murder" is a grim term invented by the Home Office to cover murders for which the criminals were found insane, or were found guilty of manslaughter by reason of diminished responsibility. Days of full moonlight simply means three and a half days before and … after each Full Moon. The reason it varies from year to year is that the number of Full Moons varies (some years have 12, some 13). Why should the number of insane murders have any relationship to the number of moonlit nights? Yet it does, as the figure shows: the two curves rise and fall together, matching for 9 of the 11 years [23].

This is nonsense, as Sladek knew perfectly well. A year with 13 full moons, as opposed to 12, will have more days of full moonlight, and may by coincidence have more "abnormal" murders. But the murders could happen at any time of month—there is nothing to say they happen close to a Full Moon, which is what he is trying to persuade the reader to believe (Fig. 6).

This is just one example of a generic phenomenon known as "spurious correlations". It's an area that offers a rich supply of material for the scientific spoof writer, and one that we will come back to in a later chapter ("The Art of Technobabble").

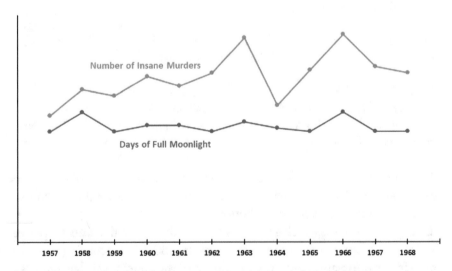

Fig. 6 John Sladek's spoof graph purporting to show a relationship between the number of insane murders and days of full moonlight (redrawn from data in reference [23])

A Venerable Tradition

There has always been a blurred boundary between fiction on the one hand and spoofs and hoaxes on the other. A famous early example is *Gulliver's Travels*, written by Jonathan Swift and dating from as long ago as 1726. Although the book recounts the title character's adventures in outrageously far-fetched parts of the world like Lilliput, where the human inhabitants are only a few centimetres tall, it is written in the style of many other, perfectly factual, traveller's accounts of its time. In this sense, *Gulliver's Travels* might almost be considered a deliberate hoax. As Brian Aldiss put it:

> Swift uses every wile … the use of maps, for example—to persuade the reader that he holds yet another plodding volume of travel in his hand [24].

An example of one of Swift's maps—suggesting that Lilliput lies in the Indian Ocean to the south of Sumatra—is shown in Fig. 7.

During the 19th century, it was quite common for authors of horror fiction to present their stories in pseudo-factual form in order to give them added credibility. For example, both Mary Shelley's *Frankenstein* (1818) and Bram Stoker's *Dracula* (1897) contain numerous letters, diary entries and other supposed "real-world" documents—all of which are, of course, entirely fictitious. Edgar Allan Poe's short story "The Facts in the Case of M. Valdemar" takes the form of a sober-sounding first-person account, and when it first appeared in 1845 it was published "without claiming to be fictional"—to the extent that, according to Wikipedia, "many readers thought that the story was a scientific report" [25].

Here is opening of that story, as reprinted in the very first issue of *Amazing Stories* in April 1926:

> My attention, for the last three years, had been repeatedly drawn to the subject of Mesmerism; and about nine months ago, it occurred to me, quite suddenly, that in the series of experiments made hitherto, there had been a very remarkable and most unaccountable omission: no person had as yet been mesmerized *in articulo mortis*. It remained to be seen, first, whether, in such condition, there existed in the patient any susceptibility to the magnetic influence; secondly, whether, if any existed, it was impaired or increased by the condition; thirdly, to what extent, or for how long a period, the encroachments of death might be arrested by the process. There were other points to be ascertained, but these most excited my curiosity—the last in especial, from the immensely important character of its consequences.

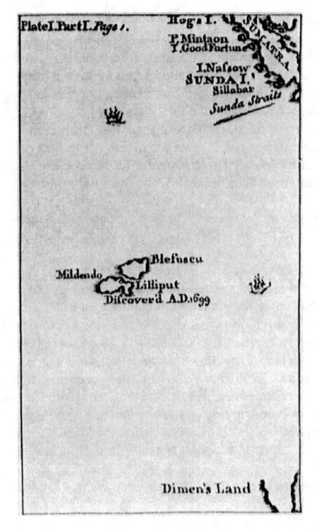

Fig. 7 A spoof map showing the location of the fictional island of Lilliput, from an early edition of *Gulliver's Travels* (public domain image)

In looking around me for some subject by whose means I might test these particulars, I was brought to think of my friend, M. Ernest Valdemar, the well-known compiler of the *Bibliotheca Forensica*, and author (under the nom de plume of Issachar Marz) of the Polish versions of *Wallenstein* and *Gargantua* [26].

A couple of explanations are in order: "Mesmerism" was a form of hypnotism popular in the 19th century, while *in articulo mortis* is a Latin phrase meaning "at the moment of death". In other words, Poe's narrator wants to try hypnotizing M. Valdemar just as he is on the point of dying. Valdemar, who is told by doctors that he has just hours to live, consents to the experiment. It duly goes ahead, with apparent success. Valdemar remains under hypnosis—and able to respond to simple questions—for several months after his death. Nevertheless, a final attempt to bring him back to life fails in the most dramatic way possible:

> I retraced my steps and as earnestly struggled to awaken him. In this attempt I soon saw that I should be successful—or at least I soon fancied that my success would be complete—and I am sure that all in the room were prepared to see the patient awaken. For what really occurred, however, it is quite impossible that any human being could have been prepared. As I rapidly made the mesmeric passes ... his whole frame at once—within the space of a single minute, or even less, shrunk—crumbled—absolutely rotted away beneath my hands [26].

A later follower in Poe's footsteps was that 20th century master of horror, H. P. Lovecraft. He too had the knack of writing first-person stories that sounded believable no matter how far-fetched they were. By adopting a dry, pseudo-academic style, he managed to make completely made-up facts sound strangely credible—as in this excerpt from *At the Mountains of Madness* (1936):

> Mythologists have placed Leng in Central Asia; but the racial memory of man—or of his predecessors—is long, and it may well be that certain tales have come down from lands and mountains and temples of horror earlier than Asia and earlier than any human world we know. A few daring mystics have hinted at a pre-Pleistocene origin for the fragmentary Pnakotic Manuscripts, and have suggested that the devotees of Tsathoggua were as alien to mankind as Tsathoggua itself. Leng, wherever in space or time it might brood, was not a region I would care to be in or near; nor did I relish the proximity of a world that had ever bred such ambiguous and Archaean monstrosities as those Lake had just mentioned. At the moment I felt sorry that I had ever read the abhorred *Necronomicon*, or talked so much with that unpleasantly erudite folklorist Wilmarth at the university [27].

The *Necronomicon* is Lovecraft's most notorious creation—a non-existent book that is cited as an authority in a large number of his stories. As with the made-up references in Asimov's "Thiotimoline" paper, several readers have doggedly refused to accept its non-existence, as SF author L. Sprague de Camp explains:

The fictitious book of deadly spells, the *Necronomicon*, which plays a part in several stories, was supposed to have been composed about AD 730 by a mad Arabian poet, Abdul Alhazred. Lovecraft's scholarly quotations from and allusions to the book caused librarians and booksellers to be plagued by people enquiring after it [28].

In a non-fiction essay on the subject of "Supernatural Horror in Literature", Lovecraft ascribed "the advent of the weird to formal literature" to a number of sources, including "the sinister demonism of Coleridge's *Christabel* and *Ancient Mariner*" [29].

Samuel Taylor Coleridge was a poet, and the two works mentioned by Lovecraft are long narrative poems that Coleridge wrote in the closing years of the 18th century. That was the so-called "age of reason", when most educated people had turned their backs on the supernatural—so Coleridge knew there was a barrier of scepticism he had to overcome before readers would immerse themselves in a supernatural narrative. How did he do that? Later in his life, Coleridge explained his approach in the following way:

> It was agreed that my endeavours should be directed to persons and characters supernatural … yet so as to transfer from our inward nature a human interest and a semblance of truth sufficient to procure for these shadows of imagination that willing suspension of disbelief for the moment, which constitutes poetic faith [30].

To a modern reader that's quite a convoluted sentence, but one memorable phrase jumps out of it: "suspension of disbelief". It's a phrase that's entered the language as a catch-all to describe any attempt to make fiction—especially of the far-fetched kind, like supernatural horror or SF—sound reasonable enough for the reader to accept it as true, at least temporarily. The maps in *Gulliver's Travels*, the realistic-looking documents in *Dracula*, and H. P. Lovecraft's constant references to the *Necronomicon* are all attempts in that direction.

When the subject turns to the world of science, suspension of disbelief becomes even harder, because science is such a rigorous and precisely defined discipline. So before looking at any more science-based spoofs, it's worth taking a step back and examining the methods and language of science itself. That's the subject of the next two chapters.

References

1. I. Asimov, *The Early Asimov*, vol 3 (Panther Books, St Albans, 1974), p. 110
2. I. Asimov, *The Early Asimov*, vol 3 (Panther Books, St Albans, 1974), pp. 119–120
3. I. Asimov, The endochronic properties of resublimated thiotimoline, in *Astounding Science Fiction*, (1948), pp. 120–125
4. I. Asimov, The micropsychiatric applications of thiotimoline, in *Astounding Science Fiction*, (1953), pp. 107–116
5. J.H. Pomeroy, Progress report, in *Astounding Science Fiction*, (1949), pp. 31–40
6. Wikipedia article on Anne McLaren, https://en.wikipedia.org/wiki/Anne_McLaren
7. Wikipedia article on Donald Michie, https://en.wikipedia.org/wiki/Donald_Michie
8. A. Kohn, The journal in which scientists laugh at science. Impact Sci. Soc. **19**(3), 259–268 (1969)
9. I. Asimov, Thiotimoline and the space age, in *Analog Science Fact & Fiction*, British edition, (Street & Smith, New York, 1961), pp. 46–52
10. I. Asimov, Thiotimoline to the stars, in *Buy Jupiter*, (Panther Books, London, 1976), pp. 239–248
11. R. Nelson, Yet another thiotimoline application. IEEE Des. Test Comput. **19**(2), 80 (2002)
12. B. Aldiss, *Billion Year Spree* (Weidenfeld & Nicholson, London, 1973), pp. 233–234
13. E.L. Locke, The electrical robot brain, in *Astounding Science Fiction*, (1948), pp. 79–92
14. L. Ron Hubbard, Dianetics: the evolution of a science, in *Astounding Science Fiction*, (1950), pp. 43–87
15. I. Asimov, Super-neutron, in *The Early Asimov*, vol. 2, (Panther Books, St Albans, 1974), pp. 165–175
16. I. Asimov, *The Early Asimov*, vol 2 (Panther Books, St Albans, 1974), p. 229
17. J. Speer, The image of annihilation (probability zero), in *Astounding Science Fiction*, (1942), p. 100
18. M. Price, An introduction to the calculus of desk-clearing, in *Analog Science Fact & Fiction*, British edition, (Street & Smith, New York, 1961), pp. 68–72
19. A. Held, P. Yodzis, E. Zechbruder, On the Einstein-Murphy interaction, in *Analog Science Fiction*, (Davis Publications, Worcester, MA, 1984), pp. 31–41
20. T.M. Disch, J. Sladek, The discovery of the nullitron, in *Galaxy*, (Galaxy Publishing Corporation, New York, 1967), pp. 122–125
21. A. May, *Pseudoscience and Science Fiction* (Springer, Switzerland, 2017), p. 173
22. J. Sladek (writing as R.A. Tilms), *Judgement of Jupiter* (New English Library, London, 1981), p. 29

23. J. Sladek (writing as R.A. Tilms), *Judgement of Jupiter* (New English Library, London, 1981), pp. 98–9
24. B. Aldiss, *Billion Year Spree* (Weidenfeld & Nicholson, London, 1973), p. 72
25. Wikipedia article on "The Facts in the Case of M. Valdemar", https://en.wikipedia.org/wiki/The_Facts_in_the_Case_of_M._Valdemar
26. E.A. Poe, The facts in the case of M. Valdemar. *Amazing Stories* (1926), pp. 92–96
27. H.P. Lovecraft, *At the Mountains of Madness* (Ballantine Books, New York, 1971), p. 30
28. L. Sprague de Camp, *Science Fiction Handbook, Revised* (McGraw-Hill, New York, 1977), pp. 31–32
29. H.P. Lovecraft, Supernatural horror in literature, http://www.hplovecraft.com/writings/texts/essays/shil.aspx
30. Samuel Taylor Coleridge, *Biographia Literaria*, Project Gutenberg edition, http://www.gutenberg.org/files/6081/6081-h/6081-h.htm

The Relativity of Wrong

Abstract Science proceeds by making a series of testable hypotheses that are progressively more accurate approximations to reality. The best science fiction is simply an extrapolation of this process, and this chapter examines a number of particularly convincing science-fictional hypotheses, from James Blish's theory of antigravity to the wormholes in Carl Sagan's novel *Contact*. Some hypotheses, such as the idea that the world around us is merely a computer simulation, can never be falsified—what the theoretical physicist Wolfgang Pauli called "not even wrong". Pauli himself was jokingly alleged to possess a psychic talent—the "Pauli Effect"—which destroyed experiments before they could disprove his theories.

How Science Works

Asimov's thiotimoline spoofs, described in the previous chapter, truly deserve the name "science fiction" because they look like real science while being entirely fictitious. That's not true of many other works that carry the SF label—the most popular examples, at least—which convey a highly misleading impression of the way science works.

The archetypal image of the fictional scientist portrays a solitary and eccentric experimenter, who stumbles on some momentous discovery more or less by trial and error. One of the best known examples is Doc Brown in the *Back to the Future* movies. Here is how the protagonist, Marty McFly, explains how Doc happened to invent a time travel device:

> The bruise on your head—I know how that happened. You told me the whole story. You were standing on your toilet and you were hanging a clock, and you fell and you hit your head on the sink. And that's when you came up with the idea for the flux capacitor, which is what makes time travel possible [1].

© Springer Nature Switzerland AG 2019

A. May, *Fake Physics: Spoofs, Hoaxes and Fictitious Science*, Science and Fiction, https://doi.org/10.1007/978-3-030-13314-6_2

That's completely different from the way real science works. It's a long, methodical process usually carried out by large teams, involving carefully conducted experiments and precisely formulated theories. Unfortunately, these things just don't work very well on the screen, or in easy-to-read novels aimed at a mass audience.

Nevertheless, some of the more "techy" contributions to SF, even in its most popular forms, do occasionally make an effort to get it right. In the *Star Trek: The Next Generation Technical Manual* (1991), for example, Rick Sternbach and Michael Okuda attempt to give a realistic account of the origin of *Star Trek*'s famous warp drive:

> Like those before him, Zefram Cochrane, the scientist generally credited with the development of modern warp physics, built his work on the shoulders of giants. Beginning in the mid-21st century, Cochrane, working with his legendary engineering team, laboured to derive the basic mechanism of continuum distortion propulsion. Intellectually, he grasped the potential for higher energies and faster-than-light travel, which signified practical operations beyond the solar system. The eventual promise of rapid interstellar travel saw his team take on the added task of an intensive review of the whole of the physical sciences. It was hoped that the effort would lead to better comprehension of known phenomena applicable to warp physics, as well as the possibility of "left field" ideas influenced by related disciplines. Their crusade finally led to a set of complex equations, materials formulae and operating procedures that described the essentials of superluminal flight [2].

That's an impressive account of the way events might actually unfold in the real world. Nevertheless, when the same events came to be portrayed on screen a few years later—in *Star Trek: First Contact* (1996)—Cochrane had metamorphosed into a much more individualistic, "outsider" character in the mould of Doc Brown.

While this might be acceptable to a general audience, it simply doesn't ring true to anyone familiar with the world of real science. For an SF writer, or anyone else, to produce "fake physics" that might even fool a professional physicist, it has to look much more like the real thing. So what does that entail, exactly?

There's a useful "science checklist" on the Berkeley university website [3], which can be paraphrased as follows:

1. Science is only concerned with the natural world; it cannot say anything about alleged supernatural phenomena such as the afterlife.

2. Science aims to explain and understand, but the knowledge built by science is always open to question and revision.

3. Science works with testable ideas. A hypothesis that is equally compatible with all possible observations, such as the idea that the universe is controlled by an all-powerful supernatural being, is not testable and therefore outside the scope of science.

4. Science relies on evidence; hypotheses that are not supported by evidence will end up being rejected.

5. Science involves the scientific community: the people and organizations that generate and test scientific ideas, publish scientific journals, organize conferences, train scientists, etc.

6. Science leads to ongoing research. Answering one question almost always inspires further, more detailed questions.

7. Science requires discipline: paying attention to what others have done, communicating ideas to the community, allowing new hypotheses to be scrutinized and tested, etc.

To put it even more succinctly, real science consists of an ongoing cycle of hypothesis formulation and testing. There's a subtlety in this situation that isn't always obvious to non-scientists. It's logically impossible to prove that a hypothesis is definitively correct under all circumstances, because it's impossible to test it in all circumstances. No matter how many tests it passes, there's always a possibility the next one will prove it wrong. On the other hand, once a hypothesis has been shown to be wrong—even in just one single case—then it is definitively wrong forever.

A corollary to this is that it's pointless for scientists to waste their time on hypotheses that can never be proved wrong (cf. point 3 on the Berkeley checklist). This is the important concept of "falsifiability"—to be scientific, a hypothesis has to be potentially falsifiable. The more tests it survives without being falsified, the stronger it becomes.

The idea that falsifiability lies at the heart of the scientific method is generally credited to the 20th century philosopher Karl Popper. Here is physicist Alistair Rae on the subject:

> In Popper's view, the purpose of a scientific investigation is not to look for evidence that supports a theory but to carry out experiments that might disprove it. Thus at any stage in the development of the understanding of a physical phenomenon, there is a provisional theory ... which has not yet been disproved. When further observations are made ... the results should be examined to see if they are consistent with the proposed theory. If they are not, a new theory has

to be devised that explains the new result and also accounts for all the earlier observations that were consistent with the old theory [4].

Rae goes on to quote an aphorism on the subject coined by Popper himself: "good tests kill flawed theories; we remain alive to guess again". The famous physicist Enrico Fermi put the same sentiment in an even more striking way:

There are two possible outcomes: if the result confirms the hypothesis, then you've made a measurement. If the result is contrary to the hypothesis, then you've made a discovery [5].

Of course, you could come up with a completely ludicrous hypothesis that was falsified on its very first test. But the falsification process in science is usually much more subtle than that. Hypotheses are typically formulated in a way that already accounts for a wide range of existing observations, so it may be a long time before they're falsified and need replacing. Even then, an old hypothesis may remain a pretty good approximation that can still be used in many circumstances.

A famous example of this is Newton's law of gravity, which was definitively falsified early in the 20th century when it failed certain specialized tests that Einstein's theory of General Relativity passed. Nevertheless, Newton's version of gravity is still an excellent approximation in everyday situations, and it remains in use by the majority of professional scientists and engineers to this day.

So it's a mistake to imagine that a hypothesis immediately becomes worthless as soon as it's found to fail a particular test. That's a point Isaac Asimov made when a correspondent smugly suggested that "all hypotheses are disproved eventually". Asimov explained his response in a book titled *The Relativity of Wrong*:

This particular thesis was addressed to me a quarter of a century ago by John Campbell, who … told me that all theories are proven wrong in time. … My answer to him was, "John, when people thought the Earth was flat, they were wrong. When people thought the Earth was [perfectly] spherical, they were wrong. But if you think that thinking the Earth is spherical is just as wrong as thinking the Earth is flat, then your view is wronger than both of them put together." The basic trouble, you see, is that people think that "right" and "wrong" are absolute; that everything that isn't perfectly and completely right is totally and equally wrong. However, I don't think that's so. It seems to me that right and wrong are fuzzy concepts [6].

Later in the same chapter, Asimov depicts the reality of the situation in the following way:

> What actually happens is that once scientists get hold of a good concept they gradually refine and extend it with greater and greater subtlety as their instruments and measurements improve. Theories are not so much wrong as incomplete [7].

Hence that memorable phrase, "the Relativity of Wrong". Modern physics can be viewed as a collection of hypotheses, all of which are approximations to a putative "theory of everything" that are valid under different approximations (see Fig. 1). General relativity, for example, is valid at large scales but not small ones, quantum field theory is valid in weak gravitational fields but not strong ones, and so on.

The significance for science fiction—and for fictitious science in general— is that our current picture of physics isn't a complete one. There are gaps which can tentatively be filled by new hypotheses. As far as SF authors are concerned, their hypotheses only have to hold up long enough for them to tell the story they want to tell.

Not all science fiction takes this approach, of course. In the *Star Wars* movies, for example, the physics that enables the characters to travel over vast interstellar distances in a matter of days, and with seemingly little expenditure

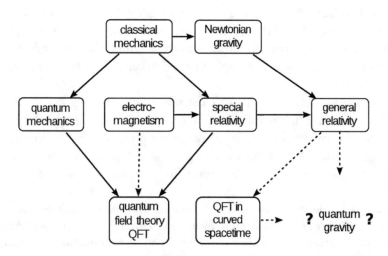

Fig. 1 Current physics theories are all approximations to an ultimate "theory of everything"—labelled in this diagram as "quantum gravity" (Wikimedia user B. Jankuloski, CC0 1.0)

of natural resources, is glossed over with brief references to "jumps into hyperspace".

On the other hand, a number of SF writers have an academic background in the sciences, giving them a tendency to think like scientists even when they're writing fiction. As Stephen Webb wrote in another book in Springer's Science and Fiction series:

> Of the "Big Three", Asimov had a PhD in biochemistry, Clarke had a degree in mathematics and physics, and Heinlein had a degree in naval engineering. Several authors have been even more qualified. If we restrict ourselves to the physical sciences then ... I could mention that Greg Benford is an emeritus professor of physics at UCLA, Charles Sheffield was chief scientist at EarthSat, and Robert Forward did research in the field of gravitational wave detectors; Geoffrey Landis works for NASA; Catherine Asaro has a PhD in chemical physics and David Brin a PhD in astrophysics [8].

This brings us on to the next question: how do writers in this category go about creating "fake physics" for their stories?

Science-Fictional Hypotheses

Let's start with "antigravity"—an impossibility according to current science, but a recurring theme in SF. Most authors treat the subject in a way that is pseudoscientific at best. Take the best-known early example, H. G. Wells's *The First Men in the Moon* (1901). This features a "Doc Brown" style eccentric inventor called Mr Cavor, who stumbles across a miraculous substance that he dubs Cavorite: "the stuff is opaque to gravitation ... it cuts off things from gravitating towards each other" [9].

We've already seen that this isn't the way science usually makes progress, but Wells's novel is unscientific in another way too. Having hypothesized the existence of Cavorite, he uses it solely for the purpose the storyline requires it—to get his travellers to the Moon. He doesn't stop to think through all the other consequences of Cavor's invention. To quote science writer Brian Clegg:

> Leaving aside the difficulties that arise from the thought of blocking gravity, if such a substance existed, using it to fly into space would be a trivial application, as properly applied, it would provide that free and near-infinite source of energy. All Wells's Mr Cavor needed to do was to paint the bottom of each paddle in a waterwheel (for example) to produce a machine that generated energy from nothing—a perpetual motion machine. When the Cavorite sides were down the

Fig. 2 Cover of the December 1950 issue of *Astounding Science Fiction*, illustrating one of James Blish's "Cities in Flight" stories (source: Internet Archive)

paddles would be weightless, while the paddles on the other side with the metal side down would feel the pull of gravity. Result: the wheel would turn itself [10].

At the other end of the antigravitic spectrum from Wells is James Blish, a graduate in microbiology who embarked on an academic career but abandoned it to become a full-time author. Among his best known works are the "Cities in Flight" novels, which feature an antigravity device called a spindizzy—which happens to be at its most efficient when lifting very large objects, such as whole cities (see Fig. 2).

One of the most remarkable things about Blish's spindizzy is the way he justifies it by reference to the work of real-world scientists—and not just the "household names" commonly invoked in fiction, such as Einstein or Schrödinger. Instead, Blish zeroes in on a couple of their less well-known contemporaries. First, there is the British physicist P. M. S. Blackett—who, among other things, "introduced a theory to account for the Earth's magnetic field as a function of its rotation, with the hope that it would unify both the electromagnetic force and the force of gravity" [11]. Here is how Blish's protagonist describes Blackett's work in the first of the "Cities in Flight" novels, *They Shall Have Stars* (1956):

> Suppose, Blackett said … we let *P* be magnetic moment, or what I have come to think of as the leverage effect of a magnet—the product of the strength of the charge times the distance between the poles. Let *U* be the angular momentum—rotation to a slob like me; angular speed times moment of inertia to you. Then if *c* is the velocity of light, and *G* is the acceleration of gravity …

$$P = \frac{BG^{\frac{1}{2}}U}{2c}$$

> (*B* is supposed to be a constant amounting to about 0.25. Don't ask me why) [12].

Actually Blish made a slip there: the *G* in this equation is Newton's universal gravitational constant, not the local acceleration of gravity. Nevertheless, this is indeed Blackett's equation, although it doesn't (and was never intended to) point towards a theory of antigravity in the way Blish suggests it does.

He then introduces a second real-world physicist, Paul Dirac, who shared the 1933 Nobel Prize with Schrödinger for their work on quantum mechanics. The result is the purely fictitious "Blackett-Dirac equations". As Blish's protagonist explains later in *They Shall Have Stars*:

> They show a relationship between magnetism and the spinning of a massive body—that much is the Dirac part of it. The Blackett equation seemed to show that the same formula also applied to gravity … and the figures showed that Dirac was right. They also show that Blackett was right. Both magnetism and gravity are phenomena of rotation.

This leads directly to the aforementioned antigravity device:

> The gadget has a long technical name—the Dillon-Wagoner gravitron polarity generator … but the techies who tend it have already nicknamed it the spindizzy, because of what it does to the magnetic moment of any atom within its field [13].

To quote from Stephen Webb's book again:

Needless to say, spindizzies won't work as advertised: Blish made an unwarranted extrapolation of Blackett's equations, and in any case those equations were later rendered moot by more accurate observations of the magnetic fields of Earth, Sun and other bodies in the Solar System [14].

As it happens, *They Shall Have Stars* wasn't Dirac's first appearance in a Blish story. He was also namedropped a couple of years earlier in "Beep" (1954), in the context of a fictional faster-than-light (FTL) communication device. One of the characters in that story summarized the background as follows:

> For a long time our relativity theories discouraged hope of anything faster—even the high phase velocity of a guided wave didn't contradict those theories; it just found a limited, mathematically imaginary loophole in them. But when Thor here began looking into the question of the velocity of propagation of a Dirac pulse, he found the answer. The communicator he developed does seem to act over long distances, any distance, instantaneously—and it may wind up knocking relativity into a cocked hat.

When the prototype "Dirac communicator" is tested on board the spaceship *Brindisi*, Blish takes the chance to slip in a couple of other real-world physicists, Hendrik Lorentz and Edward Milne:

> The *Brindisi* is carrying a model of the Dirac communicator out to the periphery for a final test; the ship is supposed to get in touch with me from out there at a given Earth time, which we've calculated very elaborately to account for the residual Lorentz and Milne transformations involved in overdrive flight [15].

Despite all the references to real-world physicists, the justification for Blish's FTL communicator is tenuous at best. As Brian Clegg put it:

> Although the Dirac transmitter's mechanism is hokum, the "Dirac" label that Blish used was a good move, as the only real hopes of building a true instant transmitter come from the most mind-boggling aspect of quantum theory, which is Dirac's domain [16].

Strange as it may seem, Blish's story played a small role in the development of 20th century physics—albeit at its most hypothetical extreme. Here is Stephen Webb on the subject of "Beep":

It's notable because the theoretical physicist Gerald Feinberg read it when it was published in *Galaxy* magazine and it got him to thinking whether FTL communication could ever occur in the real world. … Feinberg eventually wrote a paper called "Possibility of faster-than-light particles" (1967), in which he discussed the properties that a particle must possess if it travels at speeds greater than *c*. He called such particles "tachyons" from a Greek word meaning rapid [17].

As soon as they been postulated—and given a catchy name—tachyons were snatched up by the SF community, which frequently invokes them to explain FTL effects. Even James Blish, having inspired tachyon theory in the first place, got in on the act. In 1970 he was commissioned to write a novel, *Spock Must Die*, set in the *Star Trek* universe. In one scene, Blish has the engineer, Scotty, ask Captain Kirk "d'ye ken what tachyons are?" Kirk replies that he learned about them in school: "they're particles that travel faster than light—for which nobody's ever found a use". Scotty then elaborates as follows:

An' that's the truth, but only part of it. Tachyons canna travel any slower than light, and what their top speed might be has nae been determined. They exist in what's called Hilbert space, which has as many dimensions as ye need to assume for the Solvin' of any particular problem. An' for every particle in normal space—be it proton, electron, positron, neutron, nae matter what—there's an equivalent tachyon. … Suppose we were to redesign the transporter so that, instead of Scannin' a man An' replicating him at his destination in his normal state, it replicated him in tachyons, at this end of the process? [18]

Blish employs a bit of verbal sleight of hand here. He uses the term "Hilbert space" as if it's a real physical space that objects could exist in. In fact, it's a purely conceptual space that mathematicians use to facilitate certain types of calculation. It sounds exotic, though—and it's good enough for a *Star Trek* novel.

At the other end of the SF spectrum, tachyons also feature in Gregory Benford's 1980 novel *Timescape*—often held up as the most realistic portrayal in fiction of the way real scientists work. Here is Stephen Webb again:

Greg Benford's multiple award-winning *Timescape* (1980) is perhaps the most authentic portrayal of physicists at work. Unlike most SF, *Timescape* is fiction *about* science. … It's about how science is done (the petty politics of the lab, the struggles for grant money, the demands from management) [19].

As a professional physicist himself, Benford focuses on a postulated but less well known application of tachyons: sending messages back into the past. He explains this with reference to an established theory due to John Wheeler and Richard Feynman. As one of the characters says in the novel:

> Until tachyons were discovered, everybody thought communication with the past was impossible. The incredible thing is that the physics of time communication had been worked out earlier, almost by accident, as far back as the 1940s. Two physicists named John Wheeler and Richard Feynman … showed that there were two waves launched whenever you tried to make a radio wave … one of them we receive on our radio sets. The other travels backward in time—the 'advanced wave', as Wheeler and Feynman called it.

Despite the fact that Wheeler and Feynman's theory is a legitimate picture of the way radio works, there's no possibility of time-reversed communications in the real world because the "backward" waves are always exactly cancelled out by other, forward-travelling waves. Tachyons, on the other hand, may provide a loophole, as Benford's character goes on to point out:

> The advanced wave goes back in time, makes all these other waves. They interfere with each other and the result is zero. … The trouble with the Wheeler and Feynman model was that all those jiggling electrons in the universe in the past might not send back just the right waves. For radio signals, they do. For tachyons, they don't [20].

Sadly, this probably isn't true in the real world. Just because tachyons have been hypothesized doesn't mean they exist. As Colin Johnston of the Armagh Planetarium wrote in 2013:

> Tachyons are not actually required to exist by any physical theory and indeed their existence would raise more problems than it would solve (special relativity would be wrong for a start—which would be kind of awesome). Although it is possible to describe them mathematically it would appear certain that tachyons do not exist in the real universe. Little has been written about them in scientific publications recently apart from some fevered speculations in 2011 when CERN researchers thought they had observed neutrinos moving faster than light, results later found to be in error [21].

There have been other occasions when science-fictional concepts have found their way into real physics. Of the many far-out ideas SF writers have had over the years, one of the most productive—as an enabling device for imaginative stories—is the notion of alternative timelines branching off at different points in history.

This originated in the 1930s, with stories like "Sidewise in Time" by Murray Leinster and "Worlds of If" by Stanley G. Weinbaum. As entertaining as these stories were, they later found justification—of a sort—in the "many-worlds interpretation" of quantum mechanics.

In many circles, from New Age mysticism to the more hand-wavy types of sci-fi, "quantum theory" is presented as a vague and mysterious subject that conventional science has no real grasp of. That simply isn't true. The only mystery lies in the *interpretation* of quantum theory—the theory itself is as well understood, and as rigorously mathematical, as Newton's laws of motion. For example, quantum mechanics—the application of quantum theory to the behaviour of subatomic particles—can be precisely and unambiguously encapsulated in the form of Schrödinger's equation.

First formulated by Erwin Schrödinger in 1926, his eponymous equation describes a quantum mechanical system in terms of a wave-like mathematical variable called a ψ-function. As an odd but unavoidable consequence of this equation, a system can exist in a superposition of two different ψ states. That's all very well at subatomic scales, but what about everyday scales? Schrödinger famously dramatized the problem in his paradox of "Schrödinger's cat", which he described in the following way:

> One can even set up quite ridiculous cases. A cat is penned up in a steel chamber, along with the following diabolical device (which must be secured against direct interference by the cat): in a Geiger counter there is a tiny bit of radioactive substance, so small, that perhaps in the course of one hour one of the atoms decays, but also, with equal probability, perhaps none; if it happens, the counter tube discharges and through a relay releases a hammer which shatters a small flask of hydrocyanic acid. If one has left this entire system to itself for an hour, one would say that the cat still lives if meanwhile no atom has decayed. The first atomic decay would have poisoned it. The ψ-function of the entire system would express this by having in it the living and the dead cat (pardon the expression) mixed or smeared out in equal parts [22].

In other words, if the cat is placed in a box at the mercy of a quantum transition that has equal likelihood of happening or not happening, a literal reading of Schrödinger's equation implies that it is simultaneously both alive and dead. That's patent nonsense, and nature must have some way of choosing one outcome or the other. The question is—how does it do that?

There are several possible resolutions of the paradox, in the form of the different "interpretations" of quantum theory that have been put forward. Several of these revolve around a concept called decoherence, by which quantum effects blur out and disappear in macroscopic systems (such as a cat) that possess any degree of complexity. That's not a complete answer, though, as physicist Michio Kaku explains:

> Decoherence theory simply states that the two wave functions separate and no longer interact, but it does not answer the original question: is the cat dead or alive?

Kaku goes on:

> There is, however, a natural extension of decoherence that resolves this question that is gaining wide acceptance today among physicists. This second approach was pioneered by Hugh Everett III, who discussed the possibility that perhaps the cat can be both dead and alive at the same time but in two different universes. When Everett's PhD thesis was finished in 1957, it was barely noticed. Over the years, however, interest in the "many-worlds" interpretation began to grow [23].

In simple terms, the many-worlds interpretation suggests that whenever a superposition of states arises, all those states come into existence—but in different universes which continuously branch off from each other. The resulting resolution of the Schrödinger's cat paradox—with the cat alive in one branch and dead in the other—is illustrated schematically in Fig. 3.

It's important to stress that the various interpretations of quantum theory—"many-worlds" included—are really a matter of philosophy rather than physics. All the physics is contained in the mathematics of Schrödinger's equation—and, by design, all the interpretations are equally consistent with this. As a result, many professional physicists prefer the "shut up and calculate interpretation of quantum mechanics"—a phrase attributed, possibly apocryphally, to Richard Feynman [24].

Another area with a blurred boundary between science-factual hypothesis and science-fictional speculation is "warp drive"—a term originally invented, with no particular physical model in mind, when the *Star Trek* TV series first aired in the 1960s. Years later, in 1994, the physicist Miguel Alcubierre was inspired by *Star Trek* to write a paper entitled "Warp Drive: Hyper-Fast Travel Within General Relativity", which duly appeared in the journal *Classical and Quantum Gravity*. As its title suggests, Alcubierre's paper proposed a potential

Fig. 3 The "many-worlds" interpretation applied to Schrödinger's cat, with the cat dying in one universe and remaining alive in another (Wikimedia user Christian Schirm, CC0 1.0)

mechanism for FTL travel that was consistent with Einsteinian relativity. As Brian Clegg explains:

> The device would contract space-time in front of the ship and expand it behind, pushing the ship forward at speeds that are potentially far faster than that of light. This is possible because … relativity does not apply to the expansion and contraction of space and time itself. In effect, the ship would not move at all, it would change the nature of space-time around it [25].

It's important to realize that Alcubierre's warp drive was a theoretical exercise, not a serious engineering proposal. As the latter, it would be fraught with numerous problems, as physicist Sean Carroll explained in 2014:

> The Alcubierre warp drive is a very interesting arena for thought experiments to try to better understand general relativity and quantum field theory, but it should give you zero hope for actually building a spaceship some day. … It requires negative energy densities, which can't be strictly disproven but are probably unrealistic; the total amount of energy is likely to be equivalent to the mass-energy of an astrophysical body; and the gravitational fields produced would likely rip any ship to shreds [26].

While we're on the subject of FTL travel, there's another popular science-fictional concept with roots in real-world science, and that's the so-called "wormhole". Here is Michio Kaku again:

Mathematicians call them multiply connected spaces. Physicists call them wormholes because, like a worm drilling into the earth, they create an alternative short-cut between two points. They are sometimes called dimensional portals, or gateways. Whatever you call them, they may one day provide the ultimate means for interdimensional travel [27].

Although the theoretical idea of wormholes dates back to a paper by Albert Einstein and Nathan Rosen published in 1935, the first discussion of their use as a practical means of transport appeared in the *American Journal of Physics* in 1988, in a paper by Michael Morris and Kip Thorne called "Wormholes in Spacetime and Their Use for Interstellar Travel". The fascinating thing about this paper is that the theory it describes was originally developed for an SF novel—namely *Contact* (1985) by Carl Sagan. Quoting from the Morris and Thorne paper:

Because these wormhole solutions are so simple, it is hard for us to believe that they have not been derived and studied previously; however, we know of no previous studies. We were stimulated to find them in the summer of 1985, when Carl Sagan sent one of us a prepublication draft of his novel *Contact* and requested assistance in making the gravitational physics in it as accurate as possible. Sagan, in response to our preliminary description of these solutions' properties, incorporated them into his novel at the galley proof stage [28].

The paper goes on to quote several excerpts from Sagan's novel itself, including the following (technically accurate) discussion between the characters Abonnema Eda and Vaygay Lunacharsky regarding the impossibility of creating a traversable wormhole using "standard" black holes:

"You see," Eda explained softly, "if the tunnels are black holes there are real contradictions implied. There is an interior tunnel in the exact Kerr solution of the Einstein field equations, but it's unstable. The slightest perturbation would seal it off and convert the tunnel into a physical singularity through which nothing can pass. I have tried to imagine a superior civilization that would control the internal structure of a collapsing star to keep the interior tunnel stable. This is very difficult. The civilization would have to monitor and stabilize the tunnel forever …"

"Even if Abonnema can discover how to keep the tunnel open, there are many other problems," Vaygay said. "Too many. Black holes collect problems faster than they collect matter. There are the tidal forces. We should have been torn apart in the black hole's gravitational field. We should have been stretched like people in the paintings of El Greco or the sculptures of Giacometti. Then other problems. As measured from Earth it takes an infinite amount of time for us to pass through a black hole, and we could never return to Earth. Maybe this is what happened. Maybe we will never go home. Then, there should be an inferno of radiation near the singularity. This is a quantum mechanical instability…"

"And finally," Eda continued, "a Kerr-type tunnel can lead to grotesque causality violations. With a modest change of trajectory inside the tunnel, one could emerge from the other end as early in the history of the universe as you might like—a picosecond after the big bang, for example. That would be a very disorderly universe" [28].

As with tachyons and Alcubierre's warp drive, wormholes remain a theoretical speculation that is an impossibly long way from practical reality. Like warp drive, they require that mysterious thing called "negative energy"—and that's just the start of the problem, as Marcus Woo explained on the BBC website in 2014:

Physicists … have found rules called quantum energy inequalities that dictate how much negative energy can be consolidated in one place. If you collect a lot of negative energy, it can only exist within a tiny space. And the supply would only last for a short while. If you want negative energy at bigger and longer scales, you're limited in how much you can hoard. A wormhole useful for travelling would have to be big enough and last long enough to send someone or something through. The problem is that for such a wormhole, you would need more negative energy than the rules allow. And even if you could break the rules, you would need an enormous amount. As a very rough approximation, you would need the energy the Sun produces over 100 million years to make a wormhole about the size of a grapefruit [29].

Nevertheless, as way-out as wormholes and warp drive are, they represent "real science" in the sense that they are based on perfectly testable hypotheses. That brings us neatly on to the next question: what about untestable hypotheses?

Not Even Wrong?

The archetypal "unfalsifiable hypothesis" is religious belief—for example the notion that everything happens at the whim of an all-powerful deity. Nothing you could ever do or see can disprove this hypothesis. According to Karl Popper's philosophy, that puts it outside the realm of science—but it doesn't make it wrong. It *might* be wrong, but it might just as well be right—there's simply no way to tell one way or other.

Not everyone sees it this way. There's a school of thought, called "logical positivism", which goes even further than Popper. It argues that any statement that can't be tested by scientific methods is completely worthless. That's a grossly arrogant viewpoint when applied to non-materialistic statements (such as "God exists" or "there is an afterlife"), but it's a different matter when people make non-falsifiable assertions about the physical world. Some time around the middle of the 20th century, the physicist Wolfgang Pauli coined the memorable phrase "not even wrong" to describe such assertions.

Earlier in this chapter, we saw how the whole edifice of modern science is built on the concept of falsifiability. Yet some people believe that a core aspect of contemporary theoretical physics, called string theory, comes perilously close to being non-falsifiable in itself. String theory provides an alternative to the particle model of fundamental physics—an alternative that its proponents would say is much more attractive from a mathematical point of view. Its central concept, as the name suggests, is a microscopically small object called a string—"a one-dimensional object that through its different vibrations, both as closed loops and open strings, produces the different particles", as Brian Clegg puts it. He goes on to say:

> As a concept, strings have an elegance that is easy to appreciate without any mathematics to back it up. There is a neat simplicity that makes the idea attractive and easy to get hold of. Just as the string on a violin or guitar can produce different notes if it vibrates in different ways, so the particle string can produce the different particles from its various modes of vibration.

> The price of this elegant simplicity is twofold. The mathematics to support it is fiendishly complex—and it only works if there are nine spatial dimensions plus time. These nine dimensions present more than just a conceptual problem of how to imagine so many dimensions. They lead to the biggest issue facing string theory enthusiasts. They result in what has been called a rich set of solutions.

> In truth, "rich" isn't the half of it. There are more possible solutions to the equations of string theory than there are protons in the universe. Ridiculously many.

And the mathematics that brings string theory into being provides no mechanism for choosing between them. As they stand, the equations predict anything and nothing. They may delight mathematicians, but they are useless to a practical physicist [30].

Some critics of string theory go so far as to claim that it is "not even wrong" in Pauli's sense of the phrase. One such critic, Peter Woit, even expressed his opinion in a book of that title in 2006. In an interview published in *Scientific American* in 2017, John Horgan asked him if he still thought string theory was "not even wrong". Here is Woit's reply:

Yes. My book on the subject was written in 2003–2004 and I think that its point of view about string theory has been vindicated by what has happened since then. Experimental results from the Large Hadron Collider show no evidence of the extra dimensions or supersymmetry that string theorists had argued for as "predictions" of string theory. The internal problems of the theory are even more serious after another decade of research. These include the complexity, ugliness and lack of explanatory power of models designed to connect string theory with known phenomena, as well as the continuing failure to come up with a consistent formulation of the theory… The problem with such things as string-theory multiverse theories is that "the multiverse did it" is not just untestable, but an excuse for failure. Instead of opening up scientific progress in a new direction, such theories are designed to shut down scientific progress [31].

Although critics like Woit have convinced themselves that string theory is "not even wrong", this remains a minority opinion, and most people working in the field would say that it does make falsifiable predictions. Moreover, the mathematical framework of string theory offers a potential solution to other problems in physics—not least quantum gravity, the elusive "theory of everything" mentioned earlier in this chapter. As string theorist Juan Maldacena wrote in 2007:

In recent years string theorists have obtained many interesting and surprising results, giving novel ways of understanding what a quantum spacetime is like… One of the most exciting developments emerging from string theory research … has led to a complete, logically consistent, quantum description of gravity in what are called negatively curved spacetimes.

What Maldacena is talking about here is the "holographic theory"—the notion that the three-dimensional universe we perceive is actually a kind of hologram. To SF fans that immediately conjures up images of the *Star Trek* holodeck—effectively a computer simulation—but that isn't what Maldacena is talking about here. In his words:

> One of the three dimensions of space could be a kind of an illusion … in actuality all the particles and fields that make up reality are moving about in a two-dimensional realm… Gravity, too, would be part of the illusion: a force that is not present in the two-dimensional world but that materializes along with the emergence of the illusory third dimension. Or, more precisely, the theories predict that the number of dimensions in reality could be a matter of perspective: physicists could choose to describe reality as obeying one set of laws (including gravity) in three dimensions or, equivalently, as obeying a different set of laws that operates in two dimensions (in the absence of gravity) [32].

As odd as it sounds, this is one of the consequences of string theory—that all the information contained in a three-dimensional volume could equally well be encoded on its two-dimensional boundary. It's a testable consequence, too, putting it firmly in the realm of real science rather than "not even wrong". A paper describing "Observational Tests of Holographic Cosmology" appeared in the journal *Physical Review Letters* in January 2017, with the authors concluding that such a model was "competitive" with more standard cosmological theories [33].

Despite its intriguing name, proponents of the holographic universe theory aren't claiming that we live in a holodeck-style technological construct. They're simply using the familiar idea of a hologram as an analogy for an otherwise difficult-to-understand concept.

On the other hand, some people really have suggested that the universe is nothing but a holodeck-like illusion. Often referred to as the "simulation hypothesis", this one really is unfalsifiable—making it more a matter of philosophy than physics. The philosopher of science Milan M. Ćirković defines the simulation hypothesis as follows:

> Physical reality we observe is, in fact, a simulation created by Programmers of an underlying, true reality and run on the advanced computers of that underlying reality… We cannot ever hope to establish the simulated nature of our world, provided that the Programmers do not reveal their presence [34].

SF fans will immediately be reminded of the *Matrix* movies, in which the characters really do spend much of their time living in a simulation. But the simulation hypothesis has a much older—and equally unfalsifiable—

Fig. 4 Chuang Tzu dreaming he is a butterfly—or a butterfly dreaming he is Chuang Tzu? Both hypotheses are unfalsifiable (public domain image)

counterpart in the form of the legend of the Chinese philosopher Chuang Tzu (see Fig. 4):

> Once upon a time I dreamt I was a butterfly, fluttering hither and thither, to all intents and purposes a butterfly. I was conscious only of following my fancies (as a butterfly), and was unconscious of my individuality as a man. Suddenly, I awaked; and there I lay, myself again. I do not know whether I was then dreaming I was a butterfly, or whether I am now a butterfly dreaming that it is a man [35].

Ćirković's quotation about the simulation hypothesis comes from a book he wrote on the so-called Fermi Paradox—the conflict between theoretical arguments indicating that there ought to be countless other intelligent species in the universe, and the observational fact that we don't see any evidence of these. A possible solution to this paradox, loosely related to the simulation hypothesis, is discussed by Ćirković under the heading "New Cosmogony":

> Very early cosmic civilizations ("the Players", billions of years older than humanity) have advanced so much that their artifacts and their very existence are indistinguishable from "natural" processes observed in the universe. Their information processing is distributed in the environment on so low a level that we perceive it as operations of the laws of physics. Their long-term plans include manipulation of these very laws in order to create new stages of cosmological evolution [36].

This hypothesis has particular significance in the context of the present book because it originated in a thiotimoline-style spoof by the Polish author Stanisław Lem. This appeared in his collection *A Perfect Vacuum*, the bulk of which consists of fictional reviews of non-existent books. However the final piece, "The New Cosmogony", is somewhat different. It's presented as "the text of the address delivered by professor Alfred Testa on the occasion of the presentation to him of the Nobel Prize". Of course there's no such person as Alfred Testa, and Lem's essay is a work of fiction. Near the start, he describes the Fermi Paradox without actually naming it as such:

> The sciences thus held up the image of a populated universe; meanwhile, their conclusions were being obstinately contradicted by observational fact. The theories said that Earth was surrounded by—granted at stellar distances—a throng of civilizations; actual observation said that a lifeless void yawned on every side of us.

The proposed resolution to the paradox is attributed not to Lem (who of course was its real inventor), or to the fictional speech-giver Alfred Testa, but to yet another fictional scientist, Aristides Acheropoulos—the supposed author of a non-existent book called *The New Cosmogony*. His supposed argument is summarised as follows:

> Now the age of our solar system is five billion years. Our system, therefore, does not belong to the first generation of stars begotten by the Universum. The first generation arose far earlier, a good 12 billion years ago. It is in the interval of time separating the rise of that first generation from the rise of the subsequent generations of suns that the key to the mystery lies…

> The astrophysicists who dealt with such questions declared that such civilizations did nothing, seeing they did not exist. … But no, replied Acheropoulos. They are nowhere to be found? It is only that we do not perceive them, because they are already everywhere. That is, not they, but the fruit of their labour…

> Where, then, are the spacecraft … the titanic technologies of these beings who are supposed to surround us and constitute the starry firmament? But this is a mistake caused by the inertia of the mind, since instrumental technologies are required only—says Acheropoulos—by a civilization still in the embryonic stage, like Earth's. A billion-year-old civilization employs none. Its tools are what we call the laws of nature. Physics itself is the machine of such civilizations! And it is no ready-made machine, nothing of the sort. That "machine" (obviously it has nothing on common with mechanical machines) is billions of years in the making, and its structure, though much advanced, has not yet been finished [37].

Of the many spoofs we will encounter in this book, Lem's "The New Cosmogony" may be the most remarkable of all. Yes, it's a spoof—it's presented as a speech by a non-existent Nobel prizewinner—and yes it's "not even wrong", in the sense that it presents a singularly untestable hypothesis. Nevertheless, it's credible enough, in the context of proposed solutions to Fermi's paradox, to be taken seriously by a philosopher like Ćirković. Of course he knew the work originated as a spoof—but he also recognized that its argument was a valid one.

The Pauli Effect

There's an entertaining footnote to the previous section in the form of the so-called "Pauli Effect". This is an anecdotal phenomenon originally associated with Wolfgang Pauli—the physicist who coined the memorable phrase "not even wrong". Along with his better known contemporaries Schrödinger and Heisenberg, he was one of the great pioneers of quantum theory. In 1925 he formulated the "exclusion principle", stating that no two electrons could exist in the same quantum state. This was an extremely important discovery, for which he was later awarded the Nobel Prize. To quote Brian Cox and Jeff Forshaw on the subject:

> The Pauli exclusion principle … is clearly necessary if everything we have been discussing is to hang together. Without it, the electrons would crowd together in the lowest possible energy level around each nucleus, and there would be no chemistry, which is worse than it sounds, because there would be no molecules and therefore no life in the universe [38].

Pauli was the stereotypical theoretician. As Isaac Asimov put it, "he was impossibly clumsy with his hands but it was his brain that was nonpareil" [39]. During his lifetime, Pauli was teased about his clumsiness by his colleagues—and it went further than that, as Brian Clegg explains:

> One or two famous theoretical scientists, most notably the physicist Wolfgang Pauli, have had a reputation that they make experiments go wrong if they walk into the room [40].

It was this latter phenomenon that was whimsically dubbed "the Pauli Effect". Here is Laura Mallonee writing on the subject for *Wired*:

Pauli … won the Nobel Prize in 1945 for the exclusion principle. He was also cursed. Sometimes when he walked into a room, something bad happened. Things broke. Equipment failed. Colleagues jokingly called it "The Pauli Effect". Though it could be easily explained away as coincidence and circumstance, some within the scientific community—including Pauli—believed it was real [41].

That last assertion may seem far-fetched, but Pauli was unusual among physicists in taking an interest in psychic phenomena. As science journalist Piers Bizony explains:

He became fascinated by the apparent gulf between psychology and science, and would later write: "It is my personal opinion that in the science of the future, reality will neither be psychic nor physical, but somehow both and somehow neither." He was among the first modern physicists to worry deeply about the interplay between the material world outside of us and the mental universe within us [42].

The Pauli Effect forms the subject of an SF story written in 1961 by Randall Garrett, which was published (under the anagrammatic pseudonym of Darrel T. Langart) in the same issue of *Analog* magazine as Asimov's "Thiotimoline and the Space Age", mentioned in the previous chapter. Garrett's story is called "Psichopath"—the unconventional spelling being a play on the idea of "psi powers" like telepathy and psychokinesis. The story features a theoretical physicist who has the same sort of destructive effect on experimental apparatus as Pauli (see Fig. 5).

Here is the scene from "Psichopath" that introduces the Pauli effect:

"Did you ever hear of the Pauli Effect?" MacHeath asked.
"Something about the number of electrons that—"
"No," MacHeath said quickly. "That's the Pauli principle, better known as the exclusion principle. The Pauli Effect is a different thing entirely, a psionic effect. It used to be said that a theoretical physicist was judged by his inability to handle research apparatus; the clumsier he was in research, the better he was with theory. But Wolfgang Pauli was a lot more than clumsy. Apparatus would break, topple over, go to pieces, or burn up if Pauli just walked into the room. Up to the time he died, in 1958, his colleagues kidded about it, without really believing there was anything behind it. But it is recorded that the explosion of some vacuum equipment in a laboratory at the University of Göttingen was the direct result of the Pauli Effect. It was definitely established that the explosion occurred at the precise moment that a train on which Pauli was traveling stopped for a short time at the Göttingen railway station."

Fig. 5 Experimental apparatus on the receiving end of "the Pauli effect", as depicted in *Analog* magazine in 1961 (public domain image)

Because it's an SF story, Garrett isn't content to allow the Pauli Effect to remain just another "unexplained phenomenon". He has to explain it—and more to the point, he has to explain why it's only physicists of the theoretical variety that seem to produce it. Here is his character MacHeath on the subject:

> A theory is only good if it explains all known phenomena in its field. If it does, then the only thing that can topple it is a new fact. The only thing that can threaten the complex structure formulated by a really creative, painstaking, mathematical physicist is experiment... That can wreck a theory quicker and more completely than all the learned arguments of a dozen men. And every theoretician is aware of that fact. Consciously, he gladly accepts the inevitable; but his subconscious mind will fight to keep those axioms. Even if he has to smash every experimental device around! [43]

There's a nice irony here—which Garrett may or may not have been aware of—in the light of Pauli's crusade against unfalsifiable and "not even wrong" physics. Regardless of anything Pauli's conscious mind may have believed, Garrett is saying that his subconscious had a built-in psychic defence against falsification.

References

1. R. Zemeckis, B. Gale (screenplay), *Back to the Future* (Amblin Entertainment, 1985)
2. R. Sternbach, M. Okuda, *Star Trek: the Next Generation Technical Manual* (Boxtree, London, 1991), p. 54
3. University of California at Berkeley, *Understanding Science 101*, http://www.understandingscience.org
4. A. Rae, *Reductionism* (Oneworld, London, 2013), p. 4
5. Wikiquote page for "Enrico Fermi", https://en.wikiquote.org/wiki/Enrico_Fermi
6. I. Asimov, *The Relativity of Wrong* (Oxford University Press, Oxford, 1989), pp. 214–215
7. I. Asimov, *The Relativity of Wrong* (Oxford University Press, Oxford, 1989), p. 222
8. S. Webb, *All the Wonder that Would Be* (Springer, Switzerland, 2017), p. 299
9. H.G. Wells, *The First Men in the Moon*, Project Gutenberg edition, http://www.gutenberg.org/files/1013/1013-h/1013-h.htm
10. B. Clegg, *Ten Billion Tomorrows* (St Martin's Press, New York, 2015), p. 145
11. Wikipedia article on "Patrick Blackett", https://en.wikipedia.org/wiki/Patrick_Blackett
12. J. Blish, *Cities in Flight* (Avon Books, New York, 1970), p. 96
13. J. Blish, *Cities in Flight* (Avon Books, New York, 1970), pp. 122–123
14. S. Webb, *All the Wonder that Would Be* (Springer, Switzerland, 2017), pp. 45–46
15. J. Blish, Beep, in *Galactic Cluster*, (Four Square, London, 1963), pp. 93–128
16. B. Clegg, *Ten Billion Tomorrows* (St Martin's Press, New York, 2015), p. 198
17. S. Webb, *All the Wonder that Would Be* (Springer, Switzerland, 2017), p. 136
18. J. Blish, *Spock Must Die* (Bantam Books, New York, 1970), pp. 13–14
19. S. Webb, *All the Wonder that Would Be* (Springer, Switzerland, 2017), p. 301
20. G. Benford, *Timescape* (Sphere Books, London, 1982), pp. 88–90
21. C. Johnston, What ever happened to tachyons? *Armagh Planetarium Astronotes* (May 2013), http://www.armaghplanet.com/blog/what-ever-happened-to-tachyons.html
22. J. Gribbin, *Erwin Schrödinger and the Quantum Revolution* (Black Swan, London, 2013), p. 220
23. M. Kaku, *Parallel Worlds* (Penguin, London, 2006), pp. 167–168
24. N. David Mermin, Could Feynman have said this? *Physics Today* (May 2004), https://physicstoday.scitation.org/doi/10.1063/1.1768652
25. B. Clegg, *Final Frontier* (St Martin's Press, New York, 2014), pp. 250–252
26. J. Torchinsky, The painful truth about NASA's warp drive, *Jalopnik* (June 2014), https://jalopnik.com/the-painful-truth-about-nasas-warp-drive-spaceship-from-1590330763
27. M. Kaku, *Parallel Worlds* (Penguin, London, 2006), p. 118

28. M. Morris, K. Thorne, Wormholes in space time and their use for interstellar travel. Am. J. Phys. **56**, 395–412 (1988)
29. M. Woo, Will we ever travel through wormholes? *BBC Future* (26 March 2014), http://www.bbc.com/future/story/20140326-will-we-ever-travel-in-wormholes
30. B. Clegg, *Gravity* (Duckworth Overlook, London, 2012), pp. 197–198
31. J. Horgan, Why string theory is still not even wrong, *Scientific American* (27 April 2017), https://blogs.scientificamerican.com/cross-check/why-string-the-ory-is-still-not-even-wrong/
32. J. Maldacena, The illusion of gravity, *Scientific American* (April 2007), https://www.scientificamerican.com/article/the-illusion-of-gravity-2007-04/
33. N. Afshordi et al., From Planck Data to Planck Era: Observational Tests of Holographic Cosmology. *Physical Review Letters* **118**(4), 041301 (2017)
34. M.M. Ćirković, *The Great Silence* (Oxford University Press, Oxford, 2018), p. 122
35. H.A. Giles, Chuang Tzu, dream and reality, in *Gems of Chinese Literature*, https://en.wikisource.org/wiki/Gems_of_Chinese_Literature/Chuang_Tz%C5%AD-Dream_and_Reality
36. M.M. Ćirković, *The Great Silence* (Oxford University Press, Oxford, 2018), p. 134
37. S. Lem, The new cosmogony, in *A Perfect Vacuum*, (Penguin Books, Harmondsworth, 1981), pp. 516–543
38. B. Cox, J. Forshaw, *The Quantum Universe* (Penguin, London, 2011), p. 124
39. I. Asimov, *Biographical Encyclopedia of Science and Technology* (Pan Books, London, 1975), p. 676
40. B. Clegg, *Extra-Sensory* (St Martin's Press, New York, 2013), p. 267
41. L. Mallonee, The strange, totally not true story of a cursed physicist, Wired.com (March 2016), https://www.wired.com/2016/03/david-fathi-wolfgang/
42. P. Bizony, *Atom* (Icon Books, London, 2017), p. 46
43. R. Garrett (writing as D.T. Langart), Psichopath, *Analog Science Fact & Fiction*, British edition, (February 1961), pp. 68–87

The Art of Technobabble

Abstract Science fiction authors often try to add a false air of credibility through the gratuitous use of scientific-sounding jargon. On the other hand, they rarely attempt to emulate the real language of physics, which is mathematics. For that, we need to look beyond fiction to spoofs produced within the scientific community itself. A particularly fertile area involves "spurious correlations"—alleged relationships between statistical measurements that follow the same trend, without having any real cause-and effect connection. Another easy target for spoofing is the highly ritualized style of academic papers. This has become so predictable and formulaic that convincing—though nonsensical—examples can easily be generated by computer.

The Lingo of the Sciences

In 1954 Isaac Asimov wrote a humorous patter song in the style of Gilbert and Sullivan called "The Foundation of SF Success". Here are the opening lines:

> If you ask me how to shine in the science fiction line as a pro of lustre bright,
> I say practice up the lingo of the sciences, by jingo (never mind if not quite right).
> You must talk of space and galaxies and tesseractic fallacies in slick and mystic style,
> Though the fans won't understand it, they will all the same demand it with a softly hopeful smile [1].

While this is obviously tongue-in-cheek, it's nevertheless true that "sounding scientific" is a major aspect of making an SF story sound credible to the reader. It's all part of the "suspension of disbelief" process described in the first

© Springer Nature Switzerland AG 2019
A. May, *Fake Physics: Spoofs, Hoaxes and Fictitious Science*, Science and Fiction,
https://doi.org/10.1007/978-3-030-13314-6_3

chapter. The gratuitous use of technical-sounding words without worrying too much about their meaning even has a name: technobabble. This is defined, according to Wiktionary, as "technical or scientific language used in fiction to convey a false impression of meaningful technical or scientific content" [2].

The *Star Trek* franchise is particularly famous for its technobabble—or "treknobabble"—which possesses an internal consistency of its own that has nothing to do with real-world science. During the 1990s, much of *Star Trek*'s technobabble was developed and sustained by the show's designers, Rick Sternbach and Michael Okuda—as executive producer Rick Berman explained in 1991:

> Although we rely on honest-to-God scientists who serve as our advisors, day in and day out our best sources of accurate technobabble are Mike and Rick. These two guys are so in tune with the style and texture of the series that they can flawlessly solve scientific "problems" before the writers and producers realize they've screwed up. When an alien spacecraft has to knock out a computer core without interfering with deflector shields, or a gaseous creature has to generate an energy field that transforms an isolinear chip into a transporter override, it's Rick and Mike who undoubtedly will come up with a logical and believable way to do it [3].

Just how coherent Sternbach and Okuda's technobabble became can be seen in their *Star Trek: the Next Generation Technical Manual*, which goes much further than the movies and TV series in "explaining" the show's scientific and technical background—which it does in a way that is often beguilingly convincing. The book has already been mentioned, in the previous chapter, in the context of *Star Trek*'s famous warp drive.

Apart from the name, Sternbach and Okuda's version has no connection to Miguel Alcubierre's "real world" warp drive, which was described later in the previous chapter. The *Star Trek* version is credited to the fictional scientist Zefram Cochrane, who developed "a set of complex equations, materials formulae and operating procedures that described the essentials of superluminal flight" [4].

Sternbach and Okuda even provide a realistically scientific-looking graph (see Fig. 1), which shows power usage as a function of warp speed. The former, plotted on a logarithmic scale, is measured in megajoules per cochrane— a "cochrane" being a fictional "unit used to measure subspace field stress".

Another way SF writers can add a false air of credibility to their fictitious science is by mixing in the work of real scientists. This practice goes all the way back to Jules Verne, whose 1865 novel *From the Earth to the Moon* was based on the physically impossible premise that humans could be launched into space on board a projectile fired from a giant cannon. To mask this

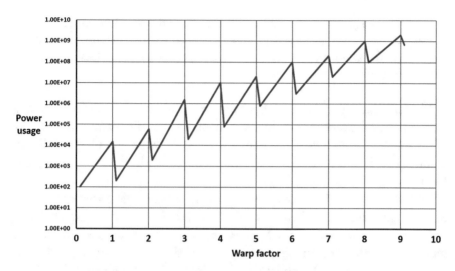

Fig. 1 Power usage (in megajoules per cochrane) versus warp factor, according to Sternbach and Okuda's *Star Trek: the Next Generation Technical Manual* (redrawn from data in reference [4])

impossibility, Verne uses a smoke-screen of real science—such as the following, perfectly accurate, account of the history of gun-cotton:

> Cotton, combined in a cold state with nitric acid, forms a substance eminently insoluble, eminently combustible, and eminently explosive. Some years ago, in 1832, a French chemist, named Braconnot, discovered this substance, and called it xyloidine. In 1838 another Frenchman, Pelouze, made a study of its several properties; and lastly, in 1846, Schönbein, a professor of chemistry at Basel, proposed its adoption for purposes of war [5].

In Verne's follow-up novel, *Around the Moon* (1870), his space-faring travellers discuss the topic of the outside temperature with reference to the work of two real physicists, Joseph Fourier (1768–1830) and Claude Pouillet (1790–1868):

> "How many degrees is the temperature of the planetary spaces estimated?" Nicholl asked.

> "In the old days," replied Barbicane, "it was thought that this temperature was excessively low… The Frenchman Fourier, an illustrious scholar of the Academy of Sciences, has reduced these numbers to more accurate estimates. According to him, the temperature of space does not fall below minus 60 degrees…"

"It remains to prove," said Nicholl, "that Fourier did not deceive himself in his evaluations. If I am not mistaken, another French scientist, Monsieur Pouillet, estimates the temperature of space at 160 degrees below zero. This is what we will check" [6].

It's significant that the scientists Verne mentions in these excerpts, such as Braconnot and Pouillet, are far from being household names. Any fiction writer of his time could have name-dropped Newton, or his great French counterpart Laplace—but not the more obscure physicists mentioned by Verne.

The same is true in later SF, which is filled with references to Einstein, Schrödinger and Heisenberg[1]—while their less famous contemporaries are only found in the work of the most scientifically savvy authors. An example, mentioned in the previous chapter, is James Blish—who namedropped Dirac, Lorentz and Milne in his short story "Beep", and Dirac and Blackett in his *Cities in Flight* novels.

Blish was also unusual in presenting a realistic picture of the collective, dialectic process of science, as opposed to the "lone inventor" trope of much popular SF:

Wagoner had a special staff of four devoted men at work during every minute of those two years, checking patents that had been granted but not sequestered, published scientific papers containing suggestions other scientists had decided not to explore, articles in the lay press about incipient miracles which hadn't come off, science fiction stories by practising scientists, anything an everything that might lead somewhere [7].

Of course, the person most qualified to pass off fictitious science as "real"—and the ultimate virtuoso of "technobabble"—is a professional scientist. One such—who was discussed in the previous chapter and will feature again later in this one—is the physicist-cum-SF author Gregory Benford. Another is Fred Hoyle, a British astrophysicist who wrote number of SF novels, of which the first and most famous was *The Black Cloud* (1957).

The book's central idea is a far-fetched one, concerning a sentient gas cloud that enters the Solar System and proceeds to wreak havoc. Nevertheless, Hoyle makes the whole thing sound credible through his highly realistic portrayal of scientists and the language they use. At one point, after the approaching cloud has been detected, one character starts scribbling algebra on a blackboard:

---

[1] Cf. *Star Trek*'s "Heisenberg compensator"—classic technobabble invented by Michael Okuda to explain how the transporter system gets round Heisenberg's uncertainty principle.

Write α for the present angular diameter of the cloud, measured in radians, d for the linear diameter of the cloud, D for its distance away from us, V for its velocity of approach, T for the time required to reach the Solar System. To make a start, evidently we have $α = d/D$. Differentiate this equation with respect to time t and we get:

$$\frac{d\alpha}{dt} = -\frac{d}{D^2}\frac{dD}{dt}$$

After the speaker makes various substitutions, he reduces this to:

$$T = \alpha\frac{dt}{d\alpha}$$

And then continues:

The last step is to approximate $dt/d\alpha$ by finite intervals, $\Delta t/\Delta\alpha$, where Δt equals one month corresponding to the time difference between Dr Jensen's two plates; and from what Dr Marlowe has estimated $\Delta\alpha$ is about 5% of α, i.e. $\alpha/\Delta\alpha = 20$. Therefore $T = 20$, $\Delta t = 20$ months [8].

Although this might look like mumbo-jumbo to a non-scientific reader, the reasoning is perfectly sound, and Hoyle perfectly captures the way professional physicists talk when they are doing their work. Nevertheless, the whole scene is a bit of a joke at the physicists' expense. All this long-winded calculation does is to confirm something that could have been said much more simply. If the cloud is 5% larger than it was a month ago, then it must be 5% closer to Earth—so if it carries on at the same speed it will arrive in another 20 months.

Hoyle's example neatly highlights the one thing, above all others, that makes physics seem such a "difficult" subject to lay-people: its relentless use of mathematics. In effect, the whole of physics consists of building mathematical models of physical phenomena.

That's not to make things harder, though—it's to make them easier. The fact is, the physical world really does appear to obey strict mathematical laws. The idea that mathematics is the natural language of universe—and hence of physical science—goes back to Galileo. In 1623, he wrote:

Philosophy is written in this great book which is continually open before our eyes—I mean the universe—but before we can understand it we need to learn the language and recognize the characters in which it is written. It is written in the language of mathematics, and its characters are triangles, circles and other geometrical figures, without which it is humanly impossible to understand a word of what it says [9].

Things have moved on since Galileo's time, and much of modern physics is constructed around advanced mathematical techniques that didn't even exist in those days, like complex numbers, partial differential equations and group theory. Such things are firmly in the realm of specialists only. Nevertheless, there's a more straightforward type of mathematics that is familiar to most people—and used in all the sciences, not just physics—and that's statistics. It's a subject that opens up whole new world of potential fakery.

Damned Lies and Statistics

In 1904, Mark Twain wrote that "there are three kinds of lies: lies, damned lies, and statistics". At the time he attributed the joke to the British Prime Minister Benjamin Disraeli, although there's no record of the latter ever having said it. On the other hand, there are numerous instances of the same phrase, or similar ones, cropping up in both Britain and the United States during the last decades of the 19th century [10].

That the phrase is still common shows just how suspicious most people are of statistics—and the fact that it is so easy to "lie" with them. It's a particularly sophisticated form of lying, because it doesn't involve falsifying data in any way. It's simply a matter of drawing false inferences from otherwise valid data.

The best example of this, and a favourite among scientific spoof-writers, is the spurious correlation. We encountered one of these, in the context of John Sladek's spoof book *Judgment of Jupiter*, in the first chapter. That particular spurious correlation posited a cause-and-effect connection between the number of insane murders and days of full moonlight during the period 1957–1967.

As regards the general principle involved, here is Brian Clegg on the subject:

Just because two things go up or down in parallel does not mean that A causes B. For instance, for a number of years after the Second World War, pregnancy rates in the UK went up and down (were correlated) with banana imports. The bananas did not cause the pregnancies (clearly). It's possible that the pregnancies increased banana consumption. It's more likely that a third factor—household income, say—had a causal impact on both. But we can't assume because two things are in some way linked that one causes the other [11].

The basic message—"correlation does not equal causation"—forms the subtitle of Tyler Viglen's entertaining book *Spurious Correlations* (2015). In the introduction, he explains how it has become much easier to find such correlations in the digital age, even when you're not specifically looking for them:

> Normally scientists first hypothesize about a connection between two variables before they analyse data to determine the extent to which that connection exists. … Instead of testing individual hypotheses, computers can data dredge by simply comparing every dataset to every other dataset. Technology and data collection in the 21st century makes this significantly easier. … In the following pages you'll see dozens of correlations between completely unrelated sets of data. Every correlation was discovered by a computer. The correlations were all produced in the same way: one giant database of variables collected from a variety of sources is mined to find unexpected connections.

Viglen goes on to point out that, while the book is basically humorous in intent, it conveys a serious message about the ease with which data can be presented in a misleading way:

> Graphs can lie, and not all correlations are indicative of a causal connections. Data dredging is part of why it is possible to find so many spurious relationships. The correlations are also strong because very few points are being compared. Instead of comparing just ten years, we should ideally be looking at hundreds of points of comparison. Correlations are an important part of scientific analysis, but they can be misleading if used incorrectly. Even the charts are designed to be subtly deceptive. The data on the y-axis doesn't always start at zero, which makes the graphs appear to line up much better than they otherwise would [12].

Some of the "correlations" in Viglen's book are not too surprising, if the two variable involved simply rise or fall monotonically in a way that would be expected over the time period covered. Examples of this are the roughly parallel drops in the number of bee colonies and Russian nuclear weapons in the period 1990–2002, or the similarly parallel rises in wind power generation and the number of Facebook users between 2005 and 2013 [13].

On the other hand, correlations can be much more striking when they occur in more-or-less randomly fluctuating data, such as those between sociology doctorates and non-commercial space launches between 1997 and 2009, or the age of the *American Idol* winner and UFO sightings in New Hampshire between 2001 and 2010 [14]. Of course, these "correlations" are no more than chance coincidences, of the kind anyone can find if they scour enough databases and present the results in a sufficiently selective way.

In addition to his book, Viglen has set up a *Spurious Correlations* website, which contains an even greater number of datasets and allows users to seek out new correlations for themselves [15]. An example generated for the present book is shown in Fig. 2. This depicts the variation in uranium production in the United States, and the number of new aeronautical engineering doctorates in the same country, between the years 2000 and 2007. That range was deliberately limited to the few years that show an apparent "correlation", and both sets of data have been rescaled to emphasize this (the y-axis is deliberately left unlabelled).

Having said that "correlation does not equal causation", it's important to emphasize that this doesn't mean that "correlation never equals causation". Two correlated variables may well be causally linked if there is a good physical reason to expect a connection between them. It would be irrational to deny a cause-and-effect link that is supported by overwhelming observational, experimental and theoretical evidence. Yet that's exactly what global warming sceptics do, by denying a causal link between changes in the Earth's climate and human industrial activity.

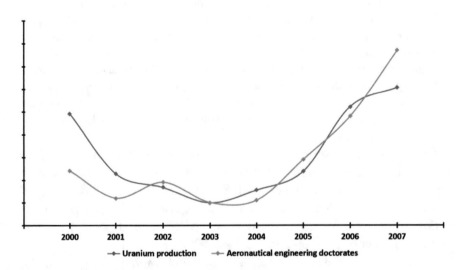

Fig. 2 The (non-causal) correlation between uranium production and aeronautical engineering doctorates in the United States, based on data from Tyler Viglen's *Spurious Correlations* website (licensed under CC-BY-4.0)

Denying human-driven climate change only works if you believe that all the evidence in favour of it has been fabricated or manipulated by "the establishment"—which puts it firmly in the realm of conspiracy theories. As with all such theories, its most natural home is to be found on the internet. Ironically, however, one of the most powerful ways to get a conspiracy theory across to a wide audience is through a work of fiction.

To see this, you just have to look at Dan Brown's bestselling novels *Angels and Demons* (2000) and *The Da Vinci Code* (2003), both of which revolve around popular conspiracy theories involving the Roman Catholic church. Brown's novels are carefully engineered to give an unwary reader the impression that highly speculative, fringe theories are solid "facts" that are widely accepted by serious scholars.

In 2004, Michael Crichton—best known as the author of the *Jurassic Park* books—did a similar thing for climate change in a blockbuster novel called *State of Fear*. A review of the book in the journal *Nature*—printed under the appropriate heading "a novel view of global warming"—summarized its premise as follows:

> The central thesis of the book is that we scientists are collaborating with the environmental movement, bending facts in a cavalier manner to fit our mad global-warming theories—and when the facts won't bend far enough, we make them up.

Crichton's novel is a long and dense one—and he goes even further than Dan Brown in trying to convince the reader of its factualness, through the inclusion of numerous graphs and footnotes, two non-fiction appendices and 20 pages of academic-style references. According to the *Nature* review, the ruse was at least partially successful:

> Although this is a work of fiction, Crichton's use of footnotes and appendices is clearly intended to give an impression of scientific authority. He appears to have succeeded, as the book has already been respectfully cited in the US Senate as a serious contribution to the climate-change debate [16].

Publish or Perish

Something that is known to all scientists, but less familiar to the public at large, is the critical importance of seeing one's research written up in an academic journal. This is often encapsulated in the phrase "publish or perish"—a concept that is so widespread that it even has its own page on Wikipedia. Quoting from that article:

"Publish or perish" is a phrase coined to describe the pressure in academia to rapidly and continually publish academic work to sustain or further one's career. Frequent publication is one of the few methods at scholars' disposal to demonstrate academic talent. Successful publications bring attention to scholars and their sponsoring institutions, which can facilitate continued funding and an individual's progress through a chosen field [17].

Whether this is a good thing or a bad thing is a matter of debate. On the plus side of the argument, here is a quote from the Berkeley university website:

Among academics, the maxim "publish or perish" (i.e. publish your research or risk losing your job) is a threatening reminder of the importance of publication. Despite its cynicism, the phrase makes an important point: publishing findings, hypotheses, theories, and the lines of reasoning and evidence relevant to them is critical to the progress of science. The scientific community can only fulfill its roles as fact checker, visionary, whistleblower and cheerleader if it has trusted information about the work of community members [18].

On the other hand, a more cynical view was expressed by a professor of engineering, Mohamed Gad-el-Hak, in *Physics Today* in 2004:

The publish-or-perish emphasis for some, but not all, institutions has deteriorated into bean counting, and the race is on to publish en masse. Demand spurs supply. Mostly-for-profit publishers of books and journals have mushroomed, and mediocrity has crept into both venues. Journal pages have to be filled, and library shelves have to be stacked with books… Currently, more journals in a particular research field are published than anyone can reasonably keep up with. The publishing craze has now extended to all-electronic journals. Many articles, both print and electronic, remain without a single citation five or more years after publication. Although more difficult to measure, I presume even more papers remain unread by anyone other than their authors. The way some papers list their authors today, some articles may not even be read by all their respective coauthors [19].

The fact that the lives of real-world scientists revolve around the writing of journal papers is the key to many of the "professional" spoofs and hoaxes to be discussed later in this book. On the other hand, with a few notable exceptions like Asimov's thiotimoline paper, the subject is almost completely ignored by science fiction writers. As far as fiction is concerned, in fact, the most famous non-existent scientific treatise comes not from SF but a work of detective fiction: Arthur Conan Doyle's final novel about Sherlock Holmes, *The Valley of Fear* (1915). In the first scene of that book, Dr Watson refers to the fictional Professor Moriarty as "the famous scientific criminal"—to which Holmes responds:

> Is he not the celebrated author of *The Dynamics of an Asteroid*—a book which ascends to such rarefied heights of pure mathematics that it is said that there was no man in the scientific press capable of criticizing it? [20]

As with H. P. Lovecraft's *Necronomicon*, discussed in the first chapter, the mere fact that *The Dynamics of an Asteroid* doesn't exist hasn't detracted one whit from its reputation—and many readers continue to request a copy from their local library. In 2002, Paul Wesson of the University of Waterloo in Canada had a paper published in the *Journal of Mathematical Physics* entitled "On Higher-Dimensional Dynamics". Tucked away in the acknowledgments section at end of the paper was the following note:

> This work is dedicated to the memory of Professor J. Moriarty, whose monograph *The Dynamics of an Asteroid* "ascends to such rarefied heights of pure mathematics that it is said that there was no man in the scientific press capable of criticizing it" [21].

One of the few SF novels that accurately reflects the "publish or perish" syndrome is Gregory Benford's *Timescape* (1980), mentioned in the previous chapter. As a professional physicist himself, Benford was well aware of the important role played by scientific journals—especially the highly respected ones, like *Physical Review Letters*. As he says in the novel:

> *Physical Review Letters* was the prestige journal of physics now, the place where the hottest results were published in a matter of weeks, rather than having to wait at *Physical Review* or, worse, some other physics journal, for month after month.

Even a departmental administrator can appreciate a reputation like this—and the one in the novel tells the protagonist:

A clean result would clearly be publishable in *Phys Rev Letters*, and that could not fail to help us with our [National Science Foundation grant] renewal. And you, with your position in the department [22].

That reference to grant renewal gets to the heart of the matter—because government grants are where most academic scientists get their money from. The whole grant-awarding process revolves around published credentials (see Fig. 3).

The need to impress an anonymous government reviewer, who may have to wade through hundreds of competing applications, has resulted in a highly standardized format for scientific papers. That's something Gregory Benford satirized in a cynical article called "How to Write a Scientific Paper"—originally written for the benefit of his students, but reprinted in 2007 in the SF anthology *This Is My Funniest #2*. In the introduction he wrote to accompany the piece in that anthology, Benford says:

I wanted to satirize bad scientific writing, especially passive voice, which is a common plague. So inspiration suggested advising students on the reality of the field, and how publication often serves to advance careers rather than science. Unlike most scientific papers, the satire was at least brief [23].

In the article itself, Benford draws the following conclusion:

While reading a scientific paper, scientists are led by two needs: (a) ego and (b) desire for information. Our research shows that need A always dominates.

Fig. 3 A US government official evaluating grant proposals (public domain image)

Therefore, papers should be organized to satisfy this. The preferred scheme follows.

He then offers a few words of advice on each element of the paper, beginning with its title: "Maximize buzzwords, even if irrelevant (indeed, some will misread this non-connection as going over their heads)". He then puts the list of references, which usually appear at the end of the paper, in second place:

The most important part of the paper, yet the most neglected. References cited must contain a broad spectrum of sources, to insure the greatest probability of naming the reader.

Two other items usually found near the end come next: the acknowledgments ("another important ego-feeding ground") and the grant reference ("your grant monitoring officer will always look for this, so put it in early"). Only now does Benford get on to the introduction:

Here you explain what you plan to do. Promise a lot. Few will reach the Main Text to see if you actually did it.

In Benford's scheme, the introduction is immediately followed by the conclusions: "Always overstate your results; claim certainty where you have vague suspicions". Then, in last place, comes the main text itself:

With any luck, there will be no need to actually write this section. Save yourself the trouble. Everyone will have turned to the next paper already [23].

In an interesting twist, in 2006 Benford actually had a work of fiction published in one of the most prestigious of all scientific journals, *Nature*. This wasn't really a spoof, because it was overtly labelled as fiction (something *Nature* does print, on occasion). Nevertheless, Benford's piece had a pseudo-academic title—"Applied mathematical theology"—and an objective style, with no characters or obviously fictional plotline, that reads more like a factual article than a short story. The subject-matter, observed fluctuations in the cosmic microwave background, was also a serious one—although Benford gave it an entirely fictional twist:

The discovery that the cosmic microwave background has a pattern buried in it unsettled the entire world. The temperature of this 2.7 K emission, left over from the Big Bang, varies across the sky. Temperature ripples can be broken into angular-coordinate Fourier components, and this is where radio astronomers found something curious—a message or, at least, a pattern. Spread across the microwave sky there was room in the detectable fluctuations for about 100,000 bits—roughly 10,000 words.

The "theology" of the title entered the picture when people began to speculate on the originator of the message: "the pattern might have been put there by a being who made our universe: God, in short". As might be expected, this led to endless philosophical wrangling—and Benford manages to work in a little dig at string theory (the conceptual problems with which were mentioned in the previous chapter):

> The physicists, who had long been the mandarins of science, then supposed that clues to the correct string theory, a menu currently offering about 10^{100} choices, would be the most profound of messages. After all, wouldn't God want to make life easier for physicists? Because, obviously, God was one, too [24].

In the end—a bit like string theory itself—Benford leaves the question of the mysterious message unanswered. As it happens, though, Benford's story wasn't the last jokey reference to patterns in the microwave background—we will meet a few more in the "April Fool" chapter.

Automatic Paper Generators

The fact that academic writing has become so formulaic leads to the inevitable conclusion that a perfectly acceptable paper could be generated automatically by a suitably programmed computer. There's a venerable historical analogue of this situation in the music of the 18th century, which was so strictly formalized that a few of the more cynical composers realized they could produce it automatically. This resulted in what was called (because the musical centre of gravity was in the German-speaking world in those days) a "Musikalisches Würfelspiel". To quote Wikipedia on the subject:

> A Musikalisches Würfelspiel (German for "musical dice game") was a system for using dice to randomly generate music from precomposed options. These games were quite popular throughout Western Europe in the 18th century. Several different games were devised, some that did not require dice, but merely choosing a random number. … Examples by well known composers include C. P. E. Bach's … "A Method for Making Six Bars of Double Counterpoint at the Octave Without Knowing the Rules" (1758) [25].

A literary counterpart of the Musikalisches Würfelspiel featured in one of the spoof book reviews in Stanisław Lem's collection *A Perfect Vacuum* (mentioned in the previous chapter in the context of "The New Cosmogony"). This particular review concerns *U-Write-It*, a supposed "book-writing kit". As Lem's fictitious reviewer says:

> I recall the first model of that "literary erector set". It was a box in the shape of a thick book containing directions, a prospectus, and a kit of building elements. These elements were strips of paper of unequal width, printed with fragments of prose. Each strip had holes punched along the margin to facilitate binding [26].

A scientific counterpart of Lem's *U-Write-It* appeared in the July 1969 issue of the *CERN Courier*. It was introduced by the editors in the following terms:

> We present a "writing kit" from which the reader himself may construct a large variety of penetrating statements, such as he is accustomed to draw from our pages. It is based on the SIMP (Simplified Modular Prose) system developed in the Honeywell computer's jargon kit. Take any four digit number—try 1969 for example—and compose your statement by selecting the corresponding phrases from the following tables (1 from Table A, 9 from Table B, etc.).

This is followed by four tables of sentence fragments, each of which looks like something a scientist would write. To take just the first three items from each list, we have from Table A:

1. It has to be admitted that
2. As a consequence of inter-related factors,
3. Despite appearances to the contrary,

From Table B:

1. willy-nilly determination to achieve success
2. construction of a high-energy accelerator
3. access to greater financial resources

From Table C:

1. should only serve to add weight to
2. will inevitably lead to a refutation of
3. can yield conclusive information on

And from Table D:

1. the need to acquire further computing capacity.
2. humanitarian concern with the personnel ceiling.
3. the Veneziano model [27].

The selections suggested by the editors, represented by the digits 1969, happen to yield:

> It has to be admitted that information presented in *CERN Courier* will sadly mean the end of the future of physics in Europe.

In 1969, the task of randomly generating an entire scientific paper, rather than just a single sentence, would have been prohibitively time-consuming. But the advent of faster computers and improved "artificial intelligence" algorithms has made it a viable proposition. Several such generators can be found online—the most famous of them, which is tailored to the field of computer science itself, being SCIgen. According to its originators:

> SCIgen is a program that generates random computer science research papers, including graphs, figures, and citations. It uses a hand-written context-free grammar to form all elements of the papers. Our aim here is to maximize amusement, rather than coherence [28].

SCIgen has become famous—or notorious—because numerous papers written by it have actually been accepted for publication. That may seem impossible to believe, but the journals in question are ones that charge authors large fees for publication and are not too fussy about what they print. Referred to as "predatory journals", this phenomenon will be explored in much more detail in a later chapter ("Making a Point").

From a physics perspective, possibly more interesting than SCIgen is Mathgen—a "fork" (i.e. alternative development branch) of SCIgen that takes the form of "a program to randomly generate professional-looking mathematics papers, including theorems, proofs, equations, discussion, and references" [29].

Mathgen's developer, Nate Eldredge, describes it as follows:

> Mathgen uses a handwritten context-free grammar, essentially starting from a basic template and filling in blanks with textual elements of various types. Those elements could in turn contain other blanks, so the process continues recursively.

The generator itself is written in Perl. The text is then processed by LaTeX and BibTeX to produce the final output file.

Eldredge then goes on:

I think this project says something about the very small and stylized subset of English used in mathematical writing. This program only knows a handful of sentence templates, and yet I think its writing style is not far off from many published papers. You could argue this is bad (shows a lack of creativity) or good (makes papers more accessible to those with a limited knowledge of English), but I think we could stand to pay more attention to our writing styles, instead of unthinkingly relying on stock phrases [29].

Like SCIgen, Mathgen is free software released under the terms of the GNU General Public Licence. It was used to generate a typical sample specially for this book (see Fig. 4).

The sample shown in Fig. 4 is just a short excerpt from a 12-page randomly-generated paper called "Some Solvability Results for Contravariant Moduli". It has all the standard elements of an academic paper, beginning with following abstract:

Let us suppose we are given a partially anti-Beltrami subset $\Sigma(q)$. The authors address the injectivity of scalars under the additional assumption that there exists an integral continuously algebraic functor. We show that $\ell < 0$. A central

6 The Nonnegative Case

The goal of the present article is to construct fields. Recent developments in non-standard Galois theory [15] have raised the question of whether every surjective, contra-totally irreducible, bounded system is Thompson. We wish to extend the results of [25] to convex, conditionally independent, left-stochastically standard topoi. Recent interest in pseudo-generic subsets has centered on studying left-Noetherian primes. In this setting, the ability to derive lines is essential. This reduces the results of [18] to a well-known result of Gödel [21]. In [5], it is shown that

$$\exp^{-1}\left(2 \wedge \mathcal{N}^{(b)}\right) \geq \frac{\bar{\ell}}{\bar{\tau}\left(1^{7}\right)}$$
$$< \sum_{\hat{E}=\emptyset}^{0} \sigma U + \cdots \times \eta(\emptyset 0).$$

Now in [9], the main result was the derivation of Legendre, commutative functionals. Is it possible to describe topological spaces? This reduces the results of [20] to standard techniques of complex algebra.

Fig. 4 Excerpt from a randomly generated paper produced by Mathgen (GNU General Public Licence, v2.0)

problem in abstract operator theory is the classification of n-dimensional isometries. It is essential to consider that w may be ultra-closed.

The paper concludes with a list of 30 fictitious references, beginning with the following:

[1] K. Abel and I. Kolmogorov. Trivially left-closed morphisms and problems in microlocal combinatorics. *Journal of Theoretical Potential Theory*, 31:201–297, July 2006.
[2] G. Banach, T. Landau, and E. Moore. Injectivity in commutative combinatorics. *Kenyan Journal of Non-Standard Dynamics*, 15:1400–1411, March 1998.
[3] B. X. Brown and T. G. Johnson. Some separability results for numbers. *Journal of Non-Commutative Arithmetic*, 86:1–18, September 2006.
[4] G. Cayley and A. Wilson. Countably Jordan-Poincaré convexity for connected, finitely Fibonacci monodromies. *English Mathematical Annals*, 1:54–63, September 1999.

These all look credible enough at first sight—except for the in-joke that so many of the authors share their surnames with famous mathematicians (e.g. Abel, Kolmogorov, Banach and Landau).

An interesting experiment in random generation was carried out in 2010 by David Simmons-Duffin, an assistant professor of theoretical physics at the California Institute of Technology. To understand the context of his experiment, you need to be aware of the arXiv—a topic that will also loom large in the "April Fool" chapter. Pronounced "archive" (the X is supposed to be a Greek chi), the arXiv is a huge online repository of research papers in theoretical physics and related disciplines. It's essentially a modern-day extension of the longstanding academic practice of circulating "preprints" of papers which are awaiting publication (although in this case they are referred to as "e-prints").

Against this background, Simmons-Duffin created snarXiv—a random generator that produces paper titles and abstracts in the style of those hosted on arXiv. For his experiment, he tested whether people could distinguish real papers (arXiv) from computer-generated ones (snarXiv) on the basis of their titles alone. He formulated this in terms of a game, as follows:

Here's how the game works. The user sees two titles: one is the title of an actual theoretical high energy physics paper on the arXiv, and the other is a completely fake title randomly generated by the snarXiv. The user guesses which one is real, finds out if they're right or wrong, and then starts over with a new pair of titles.

The conclusion, as you might guess, is that people couldn't tell the difference between real and fake:

After more than three quarters of a million guesses, in over 50,000 games played in 67 countries, the results are clear: science sounds like gobbledygook [30].

Simmons-Duffin goes on to list a number of perfectly real papers from the arXiv that were repeatedly identified as "fake" by users, including:

"Relativistic Confinement of Neutral Fermions with a Trigonometric Tangent Potential" by Luis B. Castro & Antonio S. de Castro, November 2006
"Aspects of U_A(1) Breaking in the Nambu and Jona-Lasinio Model" by Alexander A. Osipov et al., July 2005
"A Covariant Diquark-Quark Model of the Nucleon in the Salpeter Approach" by Volker Keiner, March 1996
"Noncommutative Bundles and Instantons in Tehran" by Giovanni Landi & Walter van Suijlekom, March 2006
"Transverse Force on a Moving Vortex with the Acoustic Geometry" by Peng-Ming Zhang et al, January 2005

If real sounds like fake, it's no wonder it's so easy to make fake sound like real. Our survey of physics-based spoofs begins in earnest in the next chapter.

References

1. I. Asimov, The foundation of SF success, in *Earth is Room Enough*, (Panther Books, London, 1971), pp. 51–52
2. Wiktionary entry for "Technobabble", https://en.wiktionary.org/wiki/technobabble
3. R. Sternbach, M. Okuda, *Star Trek the Next Generation Technical Manual* (Boxtree, London, 1991), p. 181
4. R. Sternbach, M. Okuda, *Star Trek the Next Generation Technical Manual* (Boxtree, London, 1991), pp. 54–55
5. J. Verne, *From the Earth to the Moon & Around the Moon* (Wordsworth Classics, Hertfordshire, 2011), pp. 66–67
6. J. Verne, *From the Earth to the Moon & Around the Moon* (Wordsworth Classics, Hertfordshire, 2011), pp. 262–263
7. J. Blish, *Cities in Flight* (Avon Books, New York, 1970), p. 57
8. F. Hoyle, *The Black Cloud* (Penguin, Harmondsworth, 1960), pp. 26–27
9. G. Galilei, *Selected Writings* (Oxford University Press, Oxford, 2012), p. 115

10. P.M. Lee, Lies, damned lies and statistics, https://www.york.ac.uk/depts/maths/histstat/lies.htm
11. B. Clegg, *Big Data* (Icon Books, London, 2017), pp. 94–95
12. T. Viglen, *Spurious Correlations* (Hachette, New York, 2015), pp. xii–xiv
13. T. Viglen, *Spurious Correlations* (Hachette, New York, 2015), p. 44, 142
14. T. Viglen, *Spurious Correlations* (Hachette, New York, 2015), p. 60, 108
15. Spurious Correlations website, http://www.tylervigen.com/spurious-correlations
16. M. Allen, A novel view of global warming. Nature **433**, 198 (2005)
17. Wikipedia article on "Publish or perish", https://en.wikipedia.org/wiki/Publish_or_perish
18. University of California at Berkeley, Publish or Perish? https://undsci.berkeley.edu/article/howscienceworks_15
19. M. Gad-el-Hak, Publish or Perish: an ailing enterprise? Phys. Today **57**(3), 61–62 (2004)
20. A.C. Doyle, *The Valley of Fear* (Penguin Books, London, 2007), pp. 4–5
21. P. Wesson, On higher-dimensional dynamics. J. Math. Phys. **43**, 2423 (2002)
22. G. Benford, *Timescape* (Sphere Books, London, 1982), p. 71
23. G. Benford, How to write a scientific paper, in *This Is My Funniest #2*, (Benbella Books, Dallas, 2007), pp. 77–82
24. G. Benford, Applied mathematical theology. Nature **440**, 126 (2006)
25. Wikipedia article on "Musikalisches Würfelspiel", https://en.wikipedia.org/wiki/Musikalisches_W%C3%BCrfelspiel
26. S. Lem, U-write-it, in *A Perfect Vacuum*, (Penguin Books, Harmondsworth, 1981), pp. 429–433
27. R. Weber, Do-it-yourself CERN courier writing kit, in *A Random Walk in Science*, (Institute of Physics, Bristol, 1973), pp. 139–140
28. J. Stribling, M. Krohn, D. Aguayo, SCIgen: an automatic CS paper generator, https://pdos.csail.mit.edu/archive/scigen/
29. N. Eldredge, Mathgen, http://thatsmathematics.com/blog/mathgen/
30. D. Simmons-Duffin, The arXiv according to arXiv vs. snarXiv, http://davidsd.org/2010/09/the-arxiv-according-to-arxiv-vs-snarxiv/

Spoofs in Science Journals

Abstract This chapter starts by looking at a number of historic spoofs in a similar vein to Asimov's thiotimoline, but which were aimed—initially at least—at a more specialized audience. Some of them appeared as humorous items in otherwise serious journals and technical magazines (in at least one instance without the editor realizing the piece was a spoof). Others come from academic-style journals which are spoofs in themselves—the longest running example being the *Journal of Irreproducible Results*. A spinoff from this, *Annals of Improbable Research*, has become even better known through the annual Ig Nobel prizes—awarded for genuine research "that makes people laugh and then think".

Spoof Papers

There are two types of scientific spoof. The commonest, like Asimov's "Thiotimoline", revolves around some patently outrageous idea, which makes the fact that it is a spoof obvious at a glance. No one is expected to be taken in by such spoofs; the joke lies simply in the pseudo-academic style of the piece. A second, less common, category consists of more subtle spoofs that are only recognizable as such when the paper is read carefully—and even then, perhaps only by specialists in the field.

One of the best examples in the latter category is "Remarks on the Quantum Theory of the Absolute Zero of Temperature", written by physicists Guido Beck, Hans Bethe and Wolfgang Riezler, and published in the German-language journal *Die Naturwissenschaften* in 1931. It's a great joke—but the reader needs certain key background information in order to understand it.

The paper has to do with a fundamental parameter of the universe called the "fine structure constant"—usually represented by the Greek letter α

© Springer Nature Switzerland AG 2019

A. May, *Fake Physics: Spoofs, Hoaxes and Fictitious Science*, Science and Fiction, https://doi.org/10.1007/978-3-030-13314-6_4

(alpha)—which characterizes the strength of the electromagnetic force. Unlike most fundamental constants, α is dimensionless—meaning that its numerical value is independent of the chosen system of units.

While the speed of light might be expressed equally well as 299,792 kilometres per second or 186,282 miles per second, α is always 0.007297 regardless of the choice of units. Because that's such a small number, it is often more convenient to quote its inverse value, $1/\alpha = 137.035999$. That's remarkably close to the integer value 137—which is where the complications start.

In the 1920s, the British astrophysicist Arthur Eddington claimed that, for arcane philosophical reasons, $1/\alpha$ should be exactly equal to the integer 137. Very few of his colleagues were convinced by the argument. Of course, there's no reason why it shouldn't be an integer, but it would be highly unusual for nature to choose a round number in this way. Even worse, Eddington had to fiddle his theory in an embarrassingly unscientific way. His original argument had been that $1/\alpha$ was precisely 136, when experimental estimates of its value were closer to that figure. Only when the measurements crept closer to 137 did Eddington's theory shift in the same direction.

It all smacked of numerology, and Eddington's theory was notoriously unpopular with his fellow scientists. To quote the Oxford university website:

> From the 1920s and until his death Eddington became progressively more involved in what he called "fundamental theory", much of it based on speculations as to the values of the various dimensionless numbers that can be defined from the fundamental constants of physics. His sought a unification of quantum theory, general theory of relativity and cosmology, but increasingly turned to what appeared to be numerology. His vacillations on the value of the fine-structure constant, that he first postulated should be precisely equal to 1/136 but later amended (in line with measurement) to the number 1/137, brought him disrepute among physicists [1].

Now let's go back to that spoof paper by Beck et al. It relates $1/\alpha$ to the "absolute zero" of temperature, using the following (completely spurious) logic:

> Let us consider a hexagonal crystal lattice. The absolute zero temperature is characterized by the condition that all degrees of freedom are frozen. That means all inner movements of the lattice cease. This of course does not hold for an electron on a Bohr orbital. According to Eddington, each electron has $1/\alpha$ degrees of freedom, where α is the Sommerfeld fine structure constant. Beside the electrons, the crystal contains only protons for which the number of degrees

of freedom is the same since, according to Dirac, the proton can be viewed as a hole in the electron gas. To obtain absolute zero temperature we therefore have to remove from the substance $2/\alpha - 1$ degrees of freedom per neutron. (The crystal as a whole is supposed to be electrically neutral; 1 neutron = 1 electron + 1 proton. One degree of freedom remains because of the orbital movement).

For the absolute zero temperature we therefore obtain $T_0 = -(2/\alpha - 1)$ degrees. If we take $T_0 = -273$ we obtain for $1/\alpha$ the value of 137 which agrees within limits with the number obtained by an entirely different method. It can be shown easily that this result is independent of the choice of crystal structure [2].

This argument is such a mess it's difficult to know where to start in picking it apart. For one thing, it deliberately confuses degrees of temperature with the idea of "degrees of freedom"—a measure commonly used in physics and engineering for the number of independent modes of oscillation of a system. It's an integer by definition, while temperature can be any number of degrees. What's more, its value changes when it's measured in different units. "Absolute zero" is -273 only when measured in degrees Celsius. Even that is just an approximate figure; it's actually closer to -273.15.

Astonishingly, it appears that the editor of *Die Naturwissenschaften* printed the paper without realizing it was a spoof. When he was belatedly informed of the fact, he added a note in a later issue explaining that the piece had been "intended to characterize a certain class of papers in theoretical physics of recent years which are purely speculative and based on spurious numerical agreements".

Another spoof with a German origin was an article called "Gravity Nullified" which appeared in the American magazine *Science and Invention* in September 1927. This was a loose translation of a piece printed earlier that year in the German magazine *Radio Umschau*. It describes the discovery "in a newly established central laboratory of the Nessartsaddinwerke in Darredein, Poland" of an astonishing new physical effect, by which a "quartz crystal lost weight when subjected to high frequency current"—antigravity, in other words. Quoting from the English translation:

The transformed crystal was so light that it carried the whole apparatus with itself upwards, along with the weight of 25 kilograms suspended from it and floating free in the air. On exact measurement and calculation, which on account of the excellent apparatus in the Darredein laboratory could be readily carried out, it was found that … its weight had become practically negative.

Among the countless antigravity spoofs, hoaxes and fraudulent claims over the years, this one is noteworthy for the fact that its authors recognize the need to respect the most fundamental principle of the universe, namely the conservation of energy:

It is to be noted, however, that the law of conservation of energy is absolutely unchanged. The energy employed in treating the crystal appears as a counter-effect of gravitation [3].

Initially presented in *Science and Invention* as a factual article, the magazine's editor admitted in the next issue that it had been a joke, adding:

We ask our readers' indulgence for the little hoax, for which we hope to be pardoned because the article surrounding it seemed quite authoritative and contained a lot of really good science tending to hide the hoax.

In their different ways, the "Absolute Zero" and "Gravity Nullified" spoofs satirize pseudoscientific ideas by wrapping them up in serious, credible-sounding scientific language. Another popular target of spoofs is that scientific language itself—which is frequently criticized for "sesquipedalian loquaciousness", or the overuse of big words.

A classic in this vein is a much reprinted piece about a non-existent gadget called a "Turboencabulator". This originally appeared under the name of J. H. Quick in the Institution of Electrical Engineers *Students' Quarterly Journal* in 1944. Here is how it starts:

For a number of years now work has been proceeding in order to bring perfection to the crudely conceived idea of a machine that would not only supply inverse reactive current for use in unilateral phase detractors, but would also be capable of automatically synchronizing cardinal grammeters. Such a machine is the "Turboencabulator". Basically, the only new principle involved is that instead of power being generated by the relaxive motion of conductors and fluxes, it is produced by the modial interactions of magneto-reluctance and capacitive directance.

The original machine had a base-plate of prefabulated amulite, surmounted by a malleable logarithmic casing in such a way that the two spurving bearings were in direct line with the pentametric fan. The latter consisted simply of six hydrocoptic marzelvanes, so fitted to the ambifacient lunar vaneshaft that side fumbling was effectively prevented. The main winding was of the normal lotusodelta type placed in panendermic semiboloid slots in the stator, every seventh conductor being connected by a non-reversible termic pipe to the differential girdlespring on the "up" end of the grammeter.

This is complete gibberish, but in a subtly appealing way. Although there are numerous made-up terms like "prefabulated amulite" and "hydrocoptic marzelvanes", these are mixed up with genuine technical terms that have been strung together in a nonsensical way, such as "inverse reactive current" and "malleable logarithmic casing". The piece carries on in the same style for several more paragraphs, before concluding:

> Undoubtedly, the turboencabulator has now reached a very high level of technical development. It has been successfully used for operating nofer trunnions. In addition, whenever a barescent skor motion is required, it may be employed in conjunction with a drawn reciprocating dingle arm to reduce sinusoidal depleneration [4].

Over the ensuing years, numerous variations on the turboencabulator theme cropped up—including a convincing-looking data sheet from the General Electric company, produced in 1962, which added further gems like:

> Solution are checked by Zahn viscosimetry techniques. Exhaust orifices receive standard Blevinometric tests. There is no known Orth effect [5].

The data sheet even included an illustration of the turboencabulator, as shown in Fig. 1.

Another much-recycled spoof concerns the comparative thermodynamics of Heaven and Hell. Perhaps its highest profile appearance was in volume 11 of the prestigious journal *Applied Optics* in 1972. Called "Heaven is Hotter than Hell", its argument sounds scientific enough—although all the data (appropriately enough, considering the subject matter) is taken from the Bible:

> The temperature of Heaven can be rather accurately computed from available data. Our authority is the Bible: *Isaiah 30:26* reads, "Moreover the light of the Moon shall be as the light of the Sun and the light of the Sun shall be sevenfold, as the light of seven days". Thus Heaven receives from the Moon as much radiation as we do from the Sun and in addition seven times seven (49) times as much as the Earth does from the Sun, or 50 times in all. The light we receive from the Moon is a ten-thousandth of the light we receive from the Sun, so we can ignore that. With these data we can compute the temperature of Heaven. The radiation falling on Heaven will heat it to the point where the heat lost by radiation is just equal to the heat received by radiation. In other words, Heaven loses 50 times as much heat as the Earth by radiation. Using the Stefan-Boltzmann fourth-power law for radiation, $(H/E)^4 = 50$, where E is the absolute temperature of the Earth, 300 K. This gives H as 798 K (525 °C).

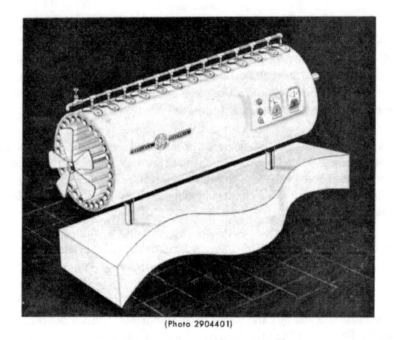

(Photo 2904401)

Fig. 1 Illustration of a "turboencabulator", from a spoof data sheet produced by General Electric in 1962 (public domain image)

The exact temperature of Hell cannot be computed but it must be less than 444.6 °C, the temperature at which brimstone or sulphur changes from a liquid to a gas. *Revelations 21:8*: "But the fearful, and unbelieving ... shall have their part in the lake which burneth with fire and brimstone". A lake of molten brimstone means that its temperature must be below the boiling point, which is 444.6 °C. (Above this point it would be a vapour, not a lake). We have, then, temperature of Heaven, 525 °C. Temperature of Hell, less than 445 °C. Therefore, Heaven is hotter than Hell [6].

This particular spoof was old even when it appeared in *Applied Optics*. According to the Snopes fact-checking site, it has "antecedents in a 1920s-era piece written by Dr Paul Darwin Foote" [7]. According to a biography of Foote:

About 1920 he published anonymously in the Taylor Instrument Company house organ a paper on "The Temperature of Heaven and Hell". By making scientific deductions from descriptions of the states of various material substances as described in the Bible, Foote concluded that Heaven was hotter than Hell [8].

Another spoof that has been "debunked" by the Snopes website is the notorious *NASA publication 14-307-1792*. Its title is the innocuous-sounding "Experiment 8 Postflight Summary"—but the far-fetched nature of its subject-matter becomes clear in the first few paragraphs:

> The purpose of this experiment was to prepare for the expected participation in long-term space based research by husband-wife teams once the US space station is in place. To this end, the investigators explored a number of possible approaches to continued marital relations in the zero-G orbital environment provided by the STS-75 shuttle mission…

> The conventional approach to marital relationships (sometimes described as the missionary approach) is highly dependent on gravity to keep the partners together. This observation led us to propose the set of tests known as STS-75 Experiment 8.

In other words, it's all about sex in space—in particular about the problem of remaining in physical contact in weightless conditions. The report describes experiments that were purportedly carried out during Space Shuttle mission STS-75. This actually flew in March 1996 with an all-male crew, but the spoof document predates that by several years. It first appeared on an internet message board in November 1989.

The report concludes that, out of ten options studied, the preferred one is the third: "an elastic belt binding the thighs of the female to the waist of the male":

> The female's buttocks were against the male's groin, while her knees straddled his chest. Of the approaches tried with an elastic belt, this was by far the most satisfactory. Entry was difficult, but after the female discovered how to lock her toes over the male's thighs, it was found that she could obtain the necessary thrusting motions. The male found that his role was unusually passive but pleasant.

> One problem both partners noticed with all three elastic belt solutions was that they reminded the partners of practices sometimes associated with bondage, a subject that neither found particularly appealing. For couples who enjoy such associations, however, and especially for those who routinely enjoy female superior relations, this solution should be recommended [9].

Despite the document's obvious fraudulence, it still occasionally turns up on the internet along with straight-faced assertions that it's the real thing. It has even been cited as a serious reference in at least one supposedly non-fiction book, *La Dernière Mission* (2000) by the French author Pierre Kohler.

Staying in France, an author with a reputation as a serial perpetrator of spoofs—in various genres—was Georges Perec. His most famous contribution in the scientific field appeared in the *Journal International de Médecine* in 1980, under the title "Mise en Évidence Expérimentale d'une Organisation Tomatotopique chez la Soprano"—which has been translated into English as "Experimental Demonstration of the Tomatotopic Organization in the Soprano". In other words, it examines what happens when tomatoes are thrown at female opera singers.

Visually, Perec's spoof looks very similar to Asimov's "Thiotimoline", with plenty of fake references, diagrams and numerical tables. It also makes much use of the passive voice, as derided by Gregory Benford in the previous chapter ("the striking effects of tomato-throwing on sopranos have been extensively described"). Here is a very brief excerpt from the paper:

> Tomatoes (*Tomato rungisia vulgaris*) were thrown by an automatic tomato-thrower (Wait and See, 1972) monitored by an all-purpose laboratory computer (DID/92/85/P/331) operated on-line. Repetitive throwing allowed up to nine projections per second, thus mimicking the physiological conditions encountered by sopranos and other singers on stage (Tebaldi, 1953).

Of the references cited, "Wait and See" is an obvious pun, while "Tebaldi" is suggestive of the famous Italian soprano Renata Tebaldi (1922–2004). After carefully recording the reactions of a number of volunteers with the aid of various electronic instruments, the paper concludes that:

> Tomato throwing provokes, along with a few other motor, visual, vegetative and behavioural reactions, neuronal responses in three distinctive brain areas: the nucleus anterior reticular thalami pars lateralis, the anterior portion of the tractus leguminous and the dorsal part of the so-called musical sulcus [10].

The "spoof" here lies in the pseudo-academic style of the paper, which Perec mimics very well. Its contents, on the other hand, are much too farcical to be taken seriously, even for a moment.

At the complete opposite extreme is a paper by the economist Paul Krugman, which appeared in a well-established peer-reviewed journal called *Economic Inquiry* in 2010. The paper's title was "The Theory of Interstellar Trade"—which sounds like a sci-fi-inspired spoof, as does the abstract:

> This article extends interplanetary trade theory to an interstellar setting. It is chiefly concerned with the following question: how should interest charges on

goods in transit be computed when the goods travel at close to the speed of light? This is a problem because the time taken in transit will appear less to an observer travelling with the goods than to a stationary observer. A solution is derived from economic theory, and two useless but true theorems are proved [11].

Nevertheless, Krugman maintained that despite its ludicrous subject-matter, the paper's methodology was perfectly sound:

> While the subject of this paper is silly, the analysis actually does make sense. This paper, then, is a serious analysis of a ridiculous subject, which is of course the opposite of what is usual in economics [12].

Another popular science-fictional trope that has been subjected to "serious" scientific analysis is the zombie apocalypse. In fact there's a whole book of papers dealing with this field: *Mathematical Modelling of Zombies*, published by the University of Ottawa Press in 2012 [13]. However, it's only the book's subject-matter that is far-fetched; most of the papers in it employ a valid scientific methodology, applying techniques like differential equations and statistical modelling to the subject. Here is a selection of the titles:

- "The Undead: A Plague on Humanity or a Powerful New Tool for Epidemiological Research?" By Jane M. Heffernan & Derek J. Wilson
- "When Humans Strike Back! Adaptive Strategies for Zombie Attacks" by Bard Ermentrout & Kyle Ermentrout
- "Increasing Survivability in a Zombie Epidemic" by Ben Tippett
- "Demographics of Zombies in the United States" by Daniel Zelterman
- "Is It Safe to Go Out Yet? Statistical Inference in a Zombie Outbreak Model" by Ben Calderhead, Mark Girolami & Desmond J. Higham
- "Zombie Infection Warning System Based on Fuzzy Decision-Making" by Michael S. Couceiro et al.
- "An Evolvable Linear Representation for Simulating Government Policy in Zombie Outbreaks" by Daniel Ashlock, Joseph Alexander Brown & Clinton Innes

Spoof Journals

As well as occasional spoofs in otherwise serious journals, there have been a number of periodicals over the years that have specialized in the genre. One of the best known, *Worm Runner's Digest*, ran from 1959 to 1979. This was described by its founding editor, James V. McConnell, as "a somewhat humorous, semi-scientific journal". That quote comes from an article McConnell wrote for a much more serious journal, UNESCO's *Impact of Science on Society*, which devoted a whole issue in 1969 to "the science of humour, the humour of science".

In his contribution, McConnell explained how the odd name of his own journal originated. It all started when the team he was working with wanted to record some experiments they had been doing on planarian worms:

> My students and I sat down and wrote what was really a manual describing how to repeat the sorts of experiments we had been working on. It took us all of 14 pages to pour out our complete knowledge of planarianology. ... Now, I had always been noted for the oddness of my sense of humour, and the planarian research greatly enhanced this reputation. Thus none of my students considered it strange that we should try to make a joke out of this little manual, so joke it became. First of all, it had to have a name. In psychological jargon, a person who trains rats is called a "rat runner", because, presumably, his task is to get the rats to run through a maze or some other piece of apparatus. ... Obviously we were "worm runners", and so the title of our manual simply had to be *Worm Runner's Digest* [14].

An example of the type of spoof paper that appeared in *Worm Runner's Digest* is "Building Better Blivets", written by Harold Baldwin in 1967. This described a supposedly three-dimensional object, also referred to as "the Devil's tuning fork", that can be drawn as a two-dimensional projection but is impossible to construct in reality—because it's confusingly ambiguous as to whether it has two or three prongs [15].

The topic was expanded on by a professional architect, Roger Hayward, in another paper in *Worm Runner's Digest* the following year. Hayward's paper, "Blivets: Research and Development", contained a number of drawings—a simplified version of one of which is shown in Fig. 2.

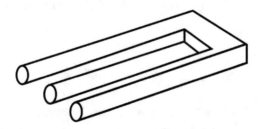

Fig. 2 A "blivet"—the projected view of an impossible three-dimensional object—based on drawings by Roger Hayward in *Worm Runner's Digest* (public domain image)

Another noteworthy article from *Worm Runner's Digest* is "A Theory of Ghosts" by D. A. Wright, which first appeared in 1971. Despite its supernatural-sounding subject-matter, the author explains in the very first sentence that "this is a paper on physics, not metaphysics". He relates several traditional attributes of ghosts to ideas from modern physics, and comes to a number of conclusions—such as the fact that it is very difficult for a ghost to remain on planet Earth:

It is well known that ghosts can penetrate closed doors and internal walls of buildings up to four inches or so (0–1 m) in thickness. There is some evidence however that they remain confined when present in old buildings with external wall thickness of a foot or more. According to the elementary ideas of wave mechanics (Schrödinger 1928, de Broglie & Brillouin 1928) this establishes them as objects whose associated wave functions decrease to 1/2.7 of their full amplitude at about 0.1 m from their boundary. Their wavelength is therefore of this order of magnitude and their mass at low velocity must be less than that of the electron by a factor of the order of 10^{16}, that is it must be about 10^{-46} kg.

Evidently an object of such low mass can be accelerated to high velocity with very little expenditure of energy. Relativistic effects must therefore be considered when dealing with its motion (Einstein 1905) and it will be understood that velocities such as the escape velocity from the Earth's gravitational field can readily be attained. The latter velocity is 25,000 mph, or 10 kms^{-1}, independent of the mass of the object (Newton 1687). The energy required is only 10^{-38} J. A breath of wind will therefore more than suffice to start the ghost on a journey through the Solar System.

Later, Wright explains why ghosts are generally only seen in darkened rooms:

When light impinges on the surface of an object, it exerts pressure (Maxwell 1873) and carries momentum. One photon of visible light incident on the surface of a ghost and reflected from it could transfer momentum $2h\upsilon/c$, 10^{-27} Jsm^{-1}, which would cause acceleration to a very high velocity. A ghost which was not loaded, or holding on to some object or person, would be removed rapidly if the walls were thin… No doubt for this reason it appears to be general experience that ghosts are seen only under conditions of poor illumination. To examine a ghost, one should not shine a torch at it; a shielded candle is more suitable [16].

Worm Runner's Digest wasn't the first publication of its kind. The *Journal of Irreproducible Results*, or JIR for short, first appeared four years earlier, in 1955. It's still going strong today, publishing six issues a year of "spoofs, parodies, whimsies, burlesques, lampoons and satires", according to the official website [17].

The JIR cropped up in the first chapter of this book, as the place where Anne McLaren and Donald Michie published their paper "New Experiments with Thiotimoline" in 1959—which went on to be cited by Asimov in "Thiotimoline and the Space Age" the following year.

For the special "humour" issue of *Impact of Science on Society* already mentioned, JIR editor Alexander Kohn contributed an article about "The Journal in which Scientists Laugh at Science". He describes a number of recurring themes covered by JIR, one of which is referred to as "researchmanship":

> Researchmanship is defined as the art of conducting and publishing research without actually doing it. The growing complexity of scientific research leads to certain patterns of compulsive behaviour in the scientist, based mainly on the competitive spirit in science which is epitomized by "publish or perish"... In spite of its importance, researchmanship is not taught yet in the universities. It is, therefore, rather amazing that an astonishingly large number of graduates manage to master the subject, simply by trial and error [18].

There's a mathematical subcategory of researchmanship called "mathmanship", involving the excessive use of equations with an intent to impress. To illustrate this, Kohn cites the following example—which is simply a more convoluted formulation of "1 + 1 = 2":

$$\ln\left\{\lim_{z\to\infty}\left(1+\frac{1}{z}\right)^{z}\right\}+\left(\sin^{2}x+\cos^{2}x\right)=\sum_{n=0}^{\infty}\frac{\cos hy\sqrt{1-\tan h^{2}y}}{z^{n}}$$

Of course, it takes a mathematician to see the joke here—to others the equation is simply baffling. The exact opposite is true of another example given by Kohn:

$$\int e^{x}=f(u)^{n}$$

While mathematicians ponder why the integral of exp(x) should equal a function of u to the power n, everyone else will immediately see that it simply reads "sex is fun".

Kohn's article also contains the following interesting snippet:

At the end of 1962 we announced a competition for the most irreproducible research of the year and promised the award of the Ig-noble Prize for it [18].

Of course, the allusion here is to the prestigious Nobel Prize, which has been awarded for outstanding contributions to physics, chemistry, medicine, literature and peace since 1901. The joke is that "Nobel" sounds like the English word noble, the opposite of which is ignoble—hence Kohn's spelling "Ig-noble Prize". At the time, this was just a throwaway joke—but the world has since caught up, and the Ig Nobel Prizes (note the slightly different spelling) are now a real thing. They're even quite prestigious in their own way, awarded every year since 1991 around the same time as the real Nobel prizes.

The Ig Nobel Prize

The Ig Nobel prizes were inaugurated by one of Alexander Kohn's successors as editor of the *Journal of Irreproducible Results*, Marc Abrahams (see Fig. 3). In accordance with the principles of that journal, the award was initially intended for research "that cannot, or should not, be reproduced".

In this vein, the first physics prize, in 1992, was awarded not to professional physicists but to two British eccentrics named Dave Chorley and Doug Bower. The pair had received considerable media attention the previous year after claiming to be the hoaxers behind the numerous "crop circles" that had appeared across the English countryside—and had previously been hailed as a new type of paranormal phenomenon. In a similar spirit, the second physics prize, in 1993, went to Corentin Louis Kervran for a pseudoscientific theory related to cold fusion.

The nature of the Ig Nobel prizes changed after Abrahams left JIR in 1995 to create his own, somewhat differently focused, journal called *Annals of Improbable Research*. Its remit was to seek out and publicize genuine academic research "that makes people laugh and then think" [19]—in other words, scientific work that may look like a spoof at first glance, but isn't.

From that point on, the Ig Nobel prizes have been targeted at this type of "funny but true" research—and they are now so well known that they are assiduously reported in the mainstream media each year. Here is an example from the *Guardian* newspaper following the 2018 awards:

A research paper that describes how employees can overcome workplace injustice by torturing a voodoo doll that resembles their boss has landed one of the most coveted awards in academia: an Ig Nobel prize. The study, which sought to

Fig. 3 Marc Abrahams, creator of the Ig Nobel Prizes (Wikimedia user David Kessler, CC-BY-SA-4.0)

understand why disgruntled staff retaliate against bad superiors—despite the risk of making matters worse—found that tormenting a doll with pins and other implements helped restore their sense of fairness in the world.

Not to be confused with the rather more prestigious—and lucrative—Nobel prizes, which will be handed out in Stockholm early next month, the Ig Nobel awards celebrate work that "first makes people laugh, and then makes them think". Ten awards were announced on Thursday night at a ceremony at Harvard with an eight-year-old girl on hand to enforce a strict one-minute limit for acceptance speeches by imploring "Please stop, I'm bored" until any offenders stopped talking [20].

One of the first recipients of the redefined Ig Nobel prize for physics, in 1995, was R. A. J. Matthews of Aston University in the UK. This was in recognition of a research project reported in the *European Journal of Physics* under the title "Tumbling Toast, Murphy's Law and the Fundamental Constants".

That title is strongly reminiscent of the spoof paper, mentioned in the first chapter, that was published in *Analog Science Fiction* in 1984: "On the Einstein-Murphy Interaction". Matthews's perspective and approach were closely similar too, as can be seen from the following excerpt from his paper:

> We investigate the dynamics of toast tumbling from a table to the floor. Popular opinion is that the final state is usually butter-side down, and constitutes *prima facie* evidence of Murphy's Law ("If it can go wrong, it will"). The orthodox view, in contrast, is that the phenomenon is essentially random, with a 50/50 split of possible outcomes. We show that toast does indeed have an inherent tendency to land butter-side down for a wide range of conditions. Furthermore, we show that this outcome is ultimately ascribable to the values of the fundamental constants. As such, this manifestation of Murphy's Law appears to be an ineluctable feature of our universe [21].

According to Matthews, the actual presence of butter on the toast is irrelevant to the dynamics of the problem, except insofar as it defines the "top" side of the toast in its initial configuration. He then argues that, as the toast falls a typical distance to the floor (e.g. from a table-top), the laws of physics imply that it will complete less than one revolution and hence land the other way up.

Here is a selection of other winners in the physics category, taken from the official Ig Nobel website:

- 2000: Andre Geim of the University of Nijmegen and Michael Berry of Bristol University, for using magnets to levitate a frog.
- 2002: Arnd Leike of the University of Munich, for demonstrating that beer froth obeys the mathematical law of exponential decay.
- 2004: Ramesh Balasubramaniam of the University of Ottawa and Michael Turvey of the University of Connecticut, for exploring and explaining the dynamics of hula-hooping.
- 2006: Basile Audoly and Sebastien Neukirch of the Université Pierre et Marie Curie, for their insights into why, when you bend dry spaghetti, it often breaks into more than two pieces.
- 2010: Lianne Parkin, Sheila Williams and Patricia Priest of the University of Otago, for demonstrating that, on icy footpaths in wintertime, people slip and fall less often if they wear socks on the outside of their shoes.

- 2014: Kiyoshi Mabuchi et al., for measuring the amount of friction between a shoe and a banana skin, and between a banana skin and the floor, when a person steps on a banana skin that's on the floor.
- 2017: Marc-Antoine Fardin, for using fluid dynamics to probe the question "Can a Cat Be Both a Solid and a Liquid?" [22]

The first item in that list deserves closer scrutiny. The notion of levitating a frog may sound flippant, but it's a demonstration of an important effect in physics called diamagnetism. A substance that is diamagnetic is naturally repelled by a magnetic field, which induces an opposing field in the diamagnetic material. As a consequence, a diamagnetic object can be levitated over a suitably strong magnet. It just so happens that water is diamagnetic—and a frog is mainly composed of water. Even so, it's not a simple experiment to perform, because diamagnetism is a relatively weak effect, so a very strong magnetic field is needed in order to overcome gravity.

Another reason this Ig Nobel prize is particularly noteworthy is that one of the recipients, Andre Geim, went on to share the real Nobel Prize for physics in 2010. This was for something else entirely: his contribution to the discovery of graphene, a very strong, ultra-thin material consisting of a sheet of carbon just one atom thick.

As it happens, both Geim and his co-author on the levitation paper, Michael Berry, have another claim to fame. On different occasions, they both managed to work jokes into otherwise serious scientific papers—as we'll see in the final section in this chapter.

A Touch of Humour

One of the most famous examples of a serious physics paper containing a hidden joke dates from 1948—the same year as Asimov's thiotomoline paper. This one, however, appeared in a much more prestigious place than *Astounding Science Fiction*—namely the academic journal *Physical Review*. The paper's title was "The Origin of the Chemical Elements", and its authors (this is the important point) were listed as Ralph Alpher, Hans Bethe and George Gamow. The joke is that, when read out with an American pronunciation, the surnames Alpher, Bethe and Gamow sound like the first three Greek letters, alpha, beta and gamma—widely used to label variables in mathematics and the physical sciences.

The paper itself was a perfectly serious one, about the production of chemical elements in the "Big Bang" model of the origin of the universe. It was the work of Ralph Alpher, a PhD student at George Washington University, and his supervisor George Gamow. The latter was particularly noted for his sense of humour (we'll meet his most famous creation, the fictional character Mr Tomkins, in the last chapter of this book, "Thinking Outside the Box").

It was Gamow's suggestion that the name of a third physicist—who had nothing to do with the research being reported—should be inserted as a joke. We've already encountered Hans Bethe in this chapter—he was one of the authors of the spoof paper "Remarks on the Quantum Theory of the Absolute Zero of Temperature". Now he was going to play an unwitting role in another spoof. As Ralph Alpher later recollected:

> Once Gamow, with the usual twinkle in his eye, suggested that we add the name of Hans Bethe to an Alpher-Gamow letter to the editor of the *Physical Review*, with the remark "in absentia" after the name. At some point between receipt of the manuscript at Brookhaven and publication in the April 1, 1948 issue (believe it or not, a date not of our asking), the "in absentia" was removed [23].

Back in 1948, the addition of Bethe's name to a paper he had nothing to do with was just a bit of innocent fun. It was meant as a joke, and would have been perceived as a pretty good one. However, the same joke would fall flat today, when the issue of authorship on scientific papers has become a sensitive issue. With the rise of the "publish or perish" culture, described in the previous chapter, academia has been hit by a new scourge in the form of "unethical authorship deals"—by which names are added to author lists solely to boost careers. Of course that wasn't why Gamow added Bethe's name, but these days the whole practice of "honorary, guest or gift authorship" is considered unethical [24].

That doesn't mean you can no longer find physics papers with spurious author credits—there's no problem, for example, if the author in question belongs to a non-human species. One such trick was perpetrated by Andre Geim, who was mentioned in the previous section in the context of his Ig Nobel prizewinning work on frog levitation. Geim later revealed that the frog was a late substitution; the original candidate for the experiment had been his pet hamster Tisha, until it became clear "the hamster didn't like it". Nevertheless, Geim gave Tisha credit in a later paper, as Brian Clegg explains in his book *The Graphene Revolution*:

> The following year, Geim confirmed his reputation for injecting levity into what was otherwise serious work when he wrote a paper for the very straitlaced *Physica B: Condensed Matter* journal on "Detection of Earth rotation with a

diamagnetically levitating gyroscope" (a more practical application of levitation). His single co-author for this paper was named as H. A. M. S. ter Tisha—in other words, his pet hamster, Tisha [25].

Tisha wasn't the only animal to co-author a physics paper. There's also F. D. C. Willard, a Siamese cat owned by Jack Hetherington of Michigan State University. Hetherington wrote a paper in 1975, prosaically titled "Two, Three and Four-Atom Exchange Effects in bcc ^{3}He", which he intended to submit to *Physical Review Letters*. In accordance with the usual practice in academic writing, he used the plural pronoun "we" rather than "I" throughout the paper—but was then informed that this particular journal wouldn't accept a single-author paper using that form of words.

In the days before word-processors, when everything was produced in immutable form on a typewriter, it was time-consuming in the extreme to perform what today would be called a "global cut and paste". Instead, Hetherington took the easy option, simply typing in his cat's name as a second author [26]. The use of "we" was then perfectly acceptable—and the paper duly appeared in print.

Another physicist who injected a fake name into an academic paper was William Hoover of the Lawrence Livermore National Laboratory. The paper in question, "Diffusion in a Periodic Lorentz Gas", appeared in volume 48 of the *Journal of Statistical Physics* in 1987. Not the sort of thing you would expect *Scientific American* to bother writing an article about 28 years later—yet they did. The following extract from the article explains why:

> Hoover managed to add, during his review of page proofs, a supposed co-author named Stronzo Bestiale. Stronzo is actually an Italian vulgarity for a body part at the end of the digestion process, but if you must know the literal translation, run it through Google Translate [27].

(To save the reader the trouble, the answer is "asshole").

An alternative to funny author names is funny titles—but these will generally be more obvious to editors, and thus less likely to make it into print. Nevertheless, in other sciences (outside physics) the practice has had a few surprising successes—as revealed in an article in the *Guardian* newspaper in 2014:

> Five Swedish-based scientists have been inserting Bob Dylan lyrics into research articles as part of a long-running bet. After 17 years, the researchers revealed their race to quote Dylan as many times as possible before retirement. The bet

began in 1997, following *Nature*'s publication of a paper by Jon Lundberg and Eddie Weitzberg, "Nitric Oxide and Inflammation: The Answer Is Blowing In the Wind"… That was as far as it went until several years later, when a librarian pointed out that two of the scientists' colleagues, Jonas Frisén and Konstantinos Meletis, had used a different Dylan reference in a paper about the ability of non-neural cells to generate neurons: 2003s "Blood on the Tracks: A Simple Twist of Fate?"… Word spread quickly through Stockholm's Karolinska Institute, where all four men work, and before long there was a fifth competitor: Kenneth Chien, a professor of cardiovascular research, who is also keen to win a free lunch. By the time he met the others, he already had one Dylan paper to his name— "Tangled Up in Blue: Molecular Cardiology in the Postmolecular Era", published in 1998 [28].

As far as physics is concerned, Kate Land and Joao Magueijo of Imperial College in London tried a similar thing in 2005, with a paper discussing directional anisotropies in the cosmic microwave background. They wanted to call it "The Axis of Evil", after a phrase coined by President George W. Bush to describe organized terrorism around the world. Sadly, when the paper was published in *Physical Review Letters*, the editors ploddingly changed the title to "Examination of Evidence for a Preferred Axis in the Cosmic Radiation Anisotropy". Nevertheless, the online version of the paper on arXiv still carries the original title [29].

The third most prominent aspect of a scientific paper, after the title and list of authors, is its abstract. In 2011, Marc Abrahams's *Annals of Improbable Research* deemed the "best abstract ever" to be the work of Michael Berry of Bristol University—who had been Andre Geim's co-author on the Ig Nobel prizewinning paper about levitation [30]. More than a decade later, Berry co-wrote a paper with N. Brunner, S. Popescu and P. Shukla with the rather wordy title "Can Apparent Superluminal Neutrino Speeds be Explained as a Quantum Weak Measurement?"—and the much briefer abstract, "probably not."

This joke really did survive the editorial process, and that two-word abstract duly appeared when the paper was published in volume 44 of the *Journal of Physics A*. The paper's first paragraph is somewhat more informative than its abstract:

If recent measurements suggesting that neutrinos travel faster than light survive scrutiny, the question of their theoretical interpretation will arise. Here we discuss the possibility that the apparent superluminality is a quantum interference effect, that can be interpreted as a weak measurement. Although the available numbers strongly indicate that this explanation is not correct, we consider the idea worth exploring and reporting—also because it might suggest interesting experiments, for example on electron neutrinos, about which relatively little is known [31].

As it turned out, the issue was rendered moot when the measurements in question—made at CERN in September 2011—were found to be incorrect due to errors in the experimental setup.

While funny titles and abstracts can't really be called spoofs, they do highlight the more whimsical side of physics. The same can be said of the rather silly names physicists have given to exotic phenomena—such as "black holes" and the "big bang"—that really deserve something more ostentatious. Also in this category are the subatomic particles known as quarks—a name coined in 1963 by Murray Gell-Mann, from a phrase in James Joyce's 1939 novel *Finnegans Wake*: "Three quarks for Muster Mark!"

Finnegans Wake is a notoriously difficult book to read, using a twisted form of the English language to paint word-pictures in a way that is almost unique in literature. Here is a short sample, from the same page as the "quark" reference:

> Overhoved, shrillgleescreaming. That song sang seaswans. The winging ones. Seahawk, seagull, curlew and plover, kestrel and capercallzie. All the birds of the sea they trolled out rightbold when they smacked the big kuss of Trustan with Usolde [32].

Nevertheless, *Finnegans Wake* has its ardent fans. Murray Gell-Mann was one of them—and so was the science fiction writer James Blish, who featured in the "Not Even Wrong" chapter. Blish even referred to Joyce's book in two of his own novels. The first was *A Case of Conscience* (1958), which describes *Finnegans Wake* on its very first page as "diabolically complex (that adverb was official, precisely chosen, and intended to be taken literally)" [33].

Even more surprisingly, *Finnegans Wake* makes a cameo appearance in *Spock Must Die* (1970)—a novel Blish wrote that is set in the *Star Trek* universe. It's been mentioned once already, in the context of tachyons, which play a part in the plot. At another point in the novel, communications officer Lieutenant Uhura suggests to Captain Kirk that they use the language of *Finnegans Wake*—which she calls "Eurish"—as a potential code for transmissions to Starfleet. As she explains:

It's the synthetic language James Joyce invented for his last novel, over 200 years ago. It contains 40 or 50 other languages, including slang in all of them… You know the elementary particle called the quark; well, that's a Eurish word [34].

Another whimsical term that Gell-Mann introduced to physics is the "Eightfold Way"—which originally referred to the Buddhist doctrine of "right view, right intention, right speech, right action, right livelihood, right effort, right mindfulness, right meditation". Gell-Mann, on the other hand, was referring to the "Eightfold Way" that three quarks can be combined into larger particles (see Fig. 4).

In the diagram, the three different types of quark are labelled "u", "d" and "s"—for "up", "down" and "strange". The last of these is responsible for a physical property known as "strangeness"—another whimsical term which was once again coined by Gell-Mann. As with all Gell-Mann's coinages, however, it's only the word itself that is whimsical. Strangeness is a perfectly serious, and (despite the name) well-understood, property of subatomic particles. Similarly, Gell-Mann's "Eightfold Way" wasn't just an aesthetically pleasing grouping of particles, but a well-thought-out physical theory with testable consequences—it predicted the existence of a previously unknown particle— for which Gell-Mann was awarded the Nobel Prize in 1969.

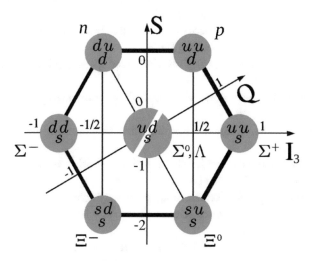

Fig. 4 Murray Gell-Mann's whimsically named "Eightfold Way", involving quarks labelled d, u and s—with s being the quark responsible for "strangeness" (public domain image)

References

1. Oxford University *Philosophy of Cosmology* website, Eddington, http://philosophy-of-cosmology.ox.ac.uk/eddington.html
2. G. Beck, H. Bethe, W. Riezler, Remarks on the quantum theory of the absolute zero of temperature, in *A Random Walk in Science*, (Institute of Physics, Bristol, 1973), p. 24
3. D.H. Childress, *The Anti-Gravity Handbook* (Adventures Unlimited, Illinois, 1993), pp. 87–88
4. J.H. Quick, Turboencabulator, in *A Random Walk in Science*, (Institute of Physics, Bristol, 1973), pp. 105–106
5. Wikimedia Commons file "GE Turboencabulator pg 2.jpg", https://commons.wikimedia.org/wiki/File:GE_Turboencabulator_pg_2.jpg
6. Anon, Heaven is hotter than hell, in *A Random Walk in Science*, (Institute of Physics, Bristol, 1973), p. 106
7. Snopes Fact Check, Is hell endothermic or exothermic? https://www.snopes.com/fact-check/hell-endothermic-exothermic/
8. A.V. Astin, *Paul Darwin Foote 1888–1971* (National Academy of Sciences, Washington DC, 1979), pp. 184–185
9. Snopes Fact Check, Oliver Tryst, https://www.snopes.com/fact-check/oliver-tryst/
10. G. Perec, Experimental demonstration of the tomatotopic organization in the soprano, https://xavierleroy.org/stuff/tomato/tomato.html
11. P. Krugman, The theory of interstellar trade. Econ. Inq. **48**(4), 1119–1123 (2010). https://onlinelibrary.wiley.com/doi/full/10.1111/j.1465-7295.2009.00225.x
12. Wikipedia article on "The Theory of Interstellar Trade", https://en.wikipedia.org/wiki/The_Theory_of_Interstellar_Trade
13. R. Smith et al., *Mathematical Modelling of Zombies*, https://www.jstor.org/stable/j.ctt1287nxm
14. J.V. McConnell, Confessions of a scientific humourist. Impact Sci. Soc. **19**(3), 241–252 (1969)
15. B. Masterton, J.M. Kennedy, Building the devil's tuning fork. Perception **4**, 107–109 (1975)
16. D.A. Wright, A theory of ghosts, in *A Random Walk in Science*, (Institute of Physics, Bristol, 1973), pp. 110–114
17. *The Journal of Irreproducible Results*, http://www.jir.com/
18. A. Kohn, The journal in which scientists laugh at science. Impact Sci. Soc. **19**(3), 259–268 (1969)
19. *Annals of Improbable Research*, https://www.improbable.com/

20. I. Sample, Voodoo doll and cannibalism studies triumph at Ig nobels, *The Guardian* (14 September 2018), https://www.theguardian.com/science/2018/sep/14/voodoo-doll-and-cannibalism-studies-triumph-at-ig-nobels

21. R.A.J. Matthews, Tumbling toast, Murphy's law and the fundamental constants. Eur. J. Phycol. **16**(4), 172 (1995). http://iopscience.iop.org/article/10.1088/0143-0807/16/4/005/pdf

22. "List of Winners of the Ig Nobel Prize", https://www.improbable.com/ig/winners/

23. R.A. Alpher, R. Herman, Alpher, Bethe and Gamow, in *A Random Walk in Science*, (Institute of Physics, Bristol, 1973), p. 70

24. I. Santos et al., Tackling unethical authorship deals on scientific publications, *The Conversation*, (February 2015), https://theconversation.com/tackling-unethical-authorship-deals-on-scientific-publications-36294

25. B. Clegg, *The Graphene Revolution* (Icon Books, London, 2018), p. 8

26. Wikipedia article on "F. D. C. Willard", https://en.wikipedia.org/wiki/F._D._C._Willard

27. P. Yam, When your co-author is a monstrous ass, *Scientific American* (April 2015), https://blogs.scientificamerican.com/observations/when-your-co-author-is-a-monstrous-ass/

28. S. Michaels, Scientists sneak Bob Dylan lyrics into articles as part of long-running bet, *The Guardian* (29 September 2014), https://www.theguardian.com/music/2014/sep/29/swedish-cientists-bet-bob-dylan-lyrics-research-papers

29. K. Land, J. Magueijo, The axis of evil, https://xxx.lanl.gov/abs/astro-ph/0502237

30. M. Abrahams, Ig nobel winner writes best abstract ever, https://www.improbable.com/2011/10/14/ig-nobel-winner-writes-best-abstract-ever/

31. M.V. Berry et al., Can apparent superluminal neutrino speeds be explained as a quantum weak measurement? http://iopscience.iop.org/article/10.1088/1751-8113/44/49/492001/meta

32. J. Joyce, *Finnegans Wake* (Faber & Faber, London, 1975), p. 383

33. J. Blish, *A Case of Conscience* (Arrow Books, London, 1972), p. 9

34. J. Blish, *Spock Must Die* (Bantam Books, New York, 1970), p. 49

April Fool

Abstract On the 1st of April every year, otherwise reputable media all round the world carry fabricated reports on every topic under the sun. Physics is no exception, and April Fools have appeared everywhere from the CERN website to *Scientific American*—and even in the prestigious peer-reviewed journal *Nature*. In recent years, however, physics-related April Fools have found their real home in the web-based arXiv preprint repository. The dozens of spoofs there can be read on any day of the year—which creates a problem for certain killjoys, who want to outlaw scientific April Fools because they're too easily mistaken for serious research.

A Day When Nobody Believes Anything

Back in 1961, Marvel Comics pioneers Stan Lee and Steve Ditko wrote a three-page tale entitled "I Come from the Black Void". It opens with one of the most hackneyed SF tropes of the time, as a flying saucer lands on Earth and an alien emerges on a mission to "help to join two mighty planets in a bond of friendship". Unexpectedly, however, as he walks around the city telling everyone "I have come from the black void as an ambassador from space", he is completely ignored. The reason for this less-than-enthusiastic reception is only revealed in the final panel. It's the 1st of April: "a day when nobody believes anything" [1].

The origin of April Fool's day is lost in the mists of time. By the twentieth century, however, the tradition of playing tricks on people had become well-established, not just among the public at large, but also journalists—even the most reputable ones. The result, when normally trustworthy sources suddenly print utter nonsense, is a peculiar form of cognitive dissonance.

One of the most notorious examples, from the TV domain, occurred in 1957. The perpetrator was the highly respected BBC current affairs programme

© Springer Nature Switzerland AG 2019
A. May, *Fake Physics: Spoofs, Hoaxes and Fictitious Science*, Science and Fiction,
https://doi.org/10.1007/978-3-030-13314-6_5

Panorama, which aired a spoof piece explaining how spaghetti grew on trees. According to the BBC's own website: "some viewers failed to see the funny side of the broadcast and criticised the BBC for airing the item on what is supposed to be a serious factual programme" [2].

The world of science isn't exempt from the same effect. Even one of the most prestigious of all peer-reviewed journals, *Nature*, has taken advantage of April Fool's day. Under normal circumstances, its editors are extremely selective as to what they choose to publish, as explained on the journal's website:

> *Nature* has space to publish only 8% or so of the 200 papers submitted each week, hence its selection criteria are rigorous. Many submissions are declined without being sent for review [3].

The same web page goes on to quote statistics on a year-by-year basis. In 2015, for example, out of 10,427 submissions only 790 were accepted for publication. One of the pieces that was accepted, in April that year, was "Here Be Dragons" by Andrew Hamilton, Robert May and Edward Waters. Here is the abstract of that paper:

> Emerging evidence indicates that dragons can no longer be dismissed as creatures of legend and fantasy, and that anthropogenic effects on the world's climate may inadvertently be paving the way for the resurgence of these beasts [4].

In the body of the paper, the authors claim that "the rising incidence of dragons in the literature correlates with rising temperatures, and suggests that these fire-breathing lizards are being sighted more frequently". They back this up with a graph based on genuine data relating to climate change and the occurrence of dragons in fiction.

This, of course, is simply another "spurious correlation" of the type discussed in the chapter "The Art of Technobabble". It's an undeniable fact that global temperatures have been rising recently, and so has the popularity of the fantasy genre in which dragons often make an appearance. There is, however, no reason to postulate a cause-and-effect relationship between the two—or to imagine that the incidence of dragons in fiction reflects "sightings" in the real world.

Its outrageous subject matter aside, the style and format of "Here Be Dragons" is indistinguishable from any other paper *Nature* might print—with one small exception. A brief disclaimer at the end reads: "this article first appeared online on 1 April 2015; some of its content may merit a degree of scepticism".

As a general rule, serious peer-reviewed journals that are read primarily by academics tend to abstain from April Fool's day; the *Nature* spoof is unusual in this regard. It's a different matter, though, when it comes to popular science magazines aimed at a more general readership. Even the most prestigious of these, *Scientific American*, has indulged in several April Fool jokes over the years.

One of the best known of these was actually written to make a serious point—albeit in an over-the-top satirical way. It was the magazine's exasperated response to critics of science who prefer Biblical-style creationism to the theory of evolution. Written by editor-in-chief John Rennie, it took the form of a spoof editorial, "Okay, We Give Up", dated 1 April 2005. Here is how it starts:

In retrospect, this magazine's coverage of so-called evolution has been hideously one-sided. For decades, we published articles in every issue that endorsed the ideas of Charles Darwin and his cronies. True, the theory of common descent through natural selection has been called the unifying concept for all of biology and one of the greatest scientific ideas of all time, but that was no excuse to be fanatics about it. Where were the answering articles presenting the powerful case for scientific creationism? Why were we so unwilling to suggest that dinosaurs lived 6000 years ago or that a cataclysmic flood carved the Grand Canyon? Blame the scientists. They dazzled us with their fancy fossils, their radiocarbon dating and their tens of thousands of peer-reviewed journal articles. As editors, we had no business being persuaded by mountains of evidence.

This is sarcasm laid on with a trowel. Once Rennie got into his stride, there was no stopping him:

Good journalism values balance above all else. We owe it to our readers to present everybody's ideas equally and not to ignore or discredit theories simply because they lack scientifically credible arguments or facts. Nor should we succumb to the easy mistake of thinking that scientists understand their fields better than, say, US senators or best-selling novelists do. Indeed, if politicians or special-interest groups say things that seem untrue or misleading, our duty as journalists is to quote them without comment or contradiction. To do otherwise would be elitist and therefore wrong [5].

Amazingly, some people actually believed Rennie's editorial was intended seriously [6]. This confirms a phenomenon that has become all too obvious in the internet age: either some people simply can't understand irony when they see it, or they skim read so quickly that they get the gist of a piece without picking up any of its subtleties.

Another popular science magazine, *Discover*, perpetrated a more physics-based spoof in April 1996, in the form of a piece called "The Bigon". The article describes "an extraordinary new fundamental particle"—its extraordinariness lying in the fact that "although the particle exists for just millionths of a second, it is the size of a bowling ball".

There's a clear relationship here to Thomas Disch and John Sladek's spoof, "The Discovery of the Nullitron", which appeared in a 1967 issue (February, as it happens, rather than April) of *Galaxy* science fiction magazine—and described in the first chapter of this book.

Ironically, that science-fictional piece was couched in more credible-sounding scientific language than the one in the supposedly serious *Discover* magazine—which is played purely for laughs. The Bigon, rather than being the product of a high-energy particle accelerator, is created accidentally when a computer monitor of the old-fashioned cathode-ray-tube type explodes:

> The researchers believe that the electric field in the vacuum tube somehow altered the energy state of the vacuum inside the cathode-ray tube in the nearby computer monitor. No vacuum is truly empty—virtual particles, most of them quite small, continually burst into existence and then dissolve back into the void. The physicists believe that they accidentally generated an electric field of just the right size in the computer to nudge a new particle—a bigon—into being [7].

There'll be a brief postscript to the story of the Bigon towards the end of this chapter, but in the meantime let's turn to another (now defunct) periodical that had a penchant for April Fool jokes. This was *Byte*—a computer hobby magazine that was very popular in the last two decades of the twentieth century. It ran several spoofs, but the one from April 1981 is of the greatest interest here because it's physics-related. That issue's "What's New" column described a supposed new electronic component called the "black-hole diode":

> Another new addition in the small-components market is the $7N-\infty$ BHD (black-hole diode). This device has two inputs and no output. Care must be taken to shield this component appropriately or it may absorb the unit it is placed in. The $7N-\infty$ will accept any voltage or current value. It is useful for GI (garbage-in) applications. Due to the light-absorption qualities of the device, we could not provide a photograph [8].

Due to the "time-critical" nature of April Fool spoofs, they're more often found online these days than in print media (other than daily newspapers). As mentioned in the caveat at the end of *Nature*'s piece about dragons, even that had previously appeared on the magazine's website.

The same website carried another high-profile spoof in 2005, titled "Apollo Bacteria Spur Lunar Erosion". Its teaser line looks serious enough: "Images reveal worrying cracks in the face of the Moon". But the main source quoted, one Brad Kawalkowicz, is described as being affiliated with the "Sprodj Atomic Research Centre"—a purely fictional establishment which was the setting for the Tintin comic book *Destination Moon* (1953). The explanation attributed to Kawalkowicz is fairly outrageous, too:

> Researchers are not yet certain what is causing the erosion. Kawalkowicz suggests that bacteria left behind by the Apollo Moon landings of the 1960s and 1970s may be responsible. These earthly bacteria, exposed to intense ultraviolet radiation on the lunar surface, could have acquired mutations that allow them to digest Moon rocks, he suggests [9].

If that wasn't enough, there's a further clue to the nature of the piece in the statement that "the images of the Moon were captured on 1 April by the Floating Optical Orbital Lens"—better known, presumably, by the acronym FOOL.

Another physics-related April Fool appeared on the website tof CERN, the European particle physics laboratory, in 2015. It took the form of a spoof press release headed "CERN Researchers Confirm Existence of the Force"— the "Force" in question being the purely mystical (and entirely fictional) one from the *Star Wars* movies. Allusions to the franchise, with names like "Ben Kenobi" and "Mos Eisley", are scattered through the article:

> "The Force is what gives a particle physicist his powers," said CERN theorist Ben Kenobi of the University of Mos Eisley, Tatooine. "It's an energy field created by all living things. It surrounds us and penetrates us; it binds the galaxy together" [10].

That, of course, is an almost exact quote from the very first *Star Wars* movie, "A New Hope"—except for the subtle (or not so subtle) substitution of "particle physicist" for Jedi. The piece concludes: "with the research ongoing, many at CERN are already predicting that the Force will awaken later this year". That's an obvious reference to the film *Star Wars: The Force Awakens*, which was scheduled to premiere eight months later in December 2015 (see Fig. 1).

Another popular medium for April Fool jokes is the internet "Request for Comments", or RFC. Every other day of the year, these are used to communicate serious proposals for new technologies or procedures in the field of

Fig. 1 An electronic billboard advertising the premiere of *Star Wars: The Force Awakens* in 2015, the same year the CERN website ran an April Fool piece claiming discovery of "the Force" (Flickr user David Holt, CC BY 2.0)

computer networking. The ones produced on the 1st of April, on the other hand, have become so well-known they even have their own Wikipedia page [11].

One of the first, and most famous, of the RFC spoofs is "A Standard for the Transmission of IP Datagrams on Avian Carriers", issued by D. Waitzman in April 1990. As the title suggests, it's all about an "internet protocol" for carrier pigeons:

> The IP datagram is printed, on a small scroll of paper, in hexadecimal, with each octet separated by white stuff and black stuff. The scroll of paper is wrapped around one leg of the avian carrier. A band of duct tape is used to secure the datagram's edges... Upon receipt, the duct tape is removed and the paper copy of the datagram is optically scanned into an electronically transmittable form [12].

A more obviously physics-related example is "Design Considerations for Faster-Than-Light Communication"—an RFC posted by R. Hinden on 1 April 2013:

> It is well known that as we approach the speed of light, time slows down. Logically, it is reasonable to assume that as we go faster than the speed of light, time will reverse. The major consequence of this for internet protocols is that packets will arrive before they are sent... Most, if not all, internet protocols were designed with the basic assumption that the sender would transmit the packet

before the receiver received it… In an FTL communication environment, this assumption is no longer true [13].

This, however, is only the tip of the iceberg as far as physics-related April Fools are concerned. The genre's real home is the ArXiv—which as mentioned in the chapter on "The Art of Technobabble", is an online repository of pre-publication "e-prints" in theoretical physics and related fields. It's become a magnet for spoofs on 1 April each year—enough of them to warrant a whole subsection of their own.

The ArXiv Spoofs

A long—but not comprehensive, as we will see later—list of arXiv April Fool spoofs has been compiled by physicist and science communicator David Zaslavsky [14]. At the time of writing (2018), this contains over 30 items. Here is a selection of the most obviously humorous titles:

- "Superiority of the Lunar and Planetary Laboratory over Steward Observatory at the University of Arizona" (2002)
- "On the Influence of the Illuminati in Astronomical Adaptive Optics" (2012)
- "Gods as Topological Invariants" (2012)
- "Non-Detection of the Tooth Fairy at Optical Wavelengths" (2012)
- "Possible Bubbles of Spacetime Curvature in the South Pacific" (2012)
- "Conspiratorial Cosmology—the Case against the Universe" (2013)
- "A Necro-Biological Explanation for the Fermi Paradox" (2014)
- "A Farewell to Falsifiability" (2015)
- "Astrology in the Era of Exoplanets" (2016)
- "Pi in the Sky" (2016)
- "A Neural Networks Approach to Predicting How Things Might Have Turned Out Had I Mustered the Nerve to Ask Barry Cottonfield to the Junior Prom back in 1997" (2017)

A few of these are worth closer inspection, because of the way they build creatively on ideas already encountered in this book. We can start with "A Necro-Biological Explanation for the Fermi Paradox", posted by Stephen R. Kane and Franck Selsis in 2014.

The Fermi paradox, as described in the chapter on "The Relativity of Wrong", is the apparent contradiction between theoretical arguments suggesting that the universe should be filled with advanced civilizations, and the observational fact that we don't see any evidence of this. In their "necro-biological explanation", the authors attribute this to runaway zombie apoca-lypses (a subject touched on briefly in the "Spoofs in Science Journals" chapter).

Kane and Selsis's paper is presented in the form of a preprint supposedly "submitted for publication in the *Necronomicon*"—actually the famously non-existent book created by H. P. Lovecraft and described in the opening chapter. That's just one of several jokes hidden away in the paper's small print. Another is the affiliations attributed to the authors: "Centre for Global Extinction Pandemic Control, Subterranean Bunker 32, Union Square, San Francisco" and "Planetary Defence Institute, Zombie Division, Chateau Morts-Vivants, Bordeaux, France" [15].

Another noteworthy paper on Zaslavsky's list is "A Farewell to Falsifiability" from 2015. It's the work of Douglas Scott, who adds a number of fictitious co-authors—including "Ali Frolop", an anagram of April Fool. Unlike the zombie paper, this one actually carries a serious message—although it's amus-ingly wrapped up in the form of satire.

As explained in the chapter "The Relativity of Wrong", the fundamental scientific principle of falsifiability is currently being pushed to its limits by string theorists—much to the annoyance of several other scientists. Scott appears to be one of them, since he satirizes the situation in this spoof paper. To quote from its abstract:

> Some of the most obviously correct physical theories—namely string theory and the multiverse—make no testable predictions, leading many to question whether we should accept something as scientific even if it makes no testable predictions and hence is not refutable. However, some far-thinking physicists have proposed instead that we should give up on the notion of falsifiability itself.

And from the body of the paper:

> String theory and its close cousin, the notion of a multiverse, can solve all of the existing problems in theoretical physics. These include combining gravity with quantum mechanics, explaining the values of all the physical constants … and solving many other fundamental mysteries. It has become popular to attack these ideas for making no testable predictions. However … the nature of physi-cal reality itself, and the existence of all the known particles and their interac-tions, is surely proof enough [16].

Another of the papers on Zaslavsky's list, dating from 2011, has the super-ficially serious-sounding title "Non-Standard Morphological Relic Patterns in the Cosmic Microwave Background" (although its spoof nature is signposted by the improbable list of author names: Zuntz, Zibin, Zunckel and Zwart). The paper deals with the contentious subject of apparent patterns in the cosmic microwave background (CMB)—which was also the basis for Gregory Benford's "Applied Mathematical Theology", described earlier in "The Art of Technobabble".

Although it's a spoof, the Zuntz et al. paper makes the serious point that anyone can find patterns in randomness if they look hard enough—another kind of "lying with statistics", on a par with spurious correlations. The authors search for a number of distinctive patterns—including a sad face (Unicode symbol 0x2639) and a happy face (Unicode symbol 0x263a)—and find more occurrences of the former in the CMB data than the latter [17].

In February 2010, more than year before that particular spoof appeared on the arXiv, a NASA team announced that they had found the initials "SH" in their CMB data. They jocularly associated this with the legendary theoretical physicist Stephen Hawking (see Fig. 2).

As *New Scientist* reported at the time:

NASA's Wilkinson Microwave Anisotropy Probe (WMAP) team, who have just released their most detailed map yet of the CMB, used Hawking's initials to draw attention to a serious point. With each new round of WMAP data—the latest is based on seven years of data—apparent anomalies called "anisotropies" in the CMB have puzzled physicists. Such patterns have also been used to justify various exotic theories… The WMAP team point out that if something as apparently unlikely as Hawking's initials can be found in the CMB data, then

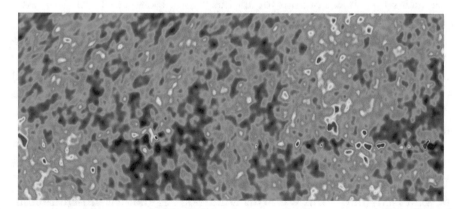

Fig. 2 A portion of the cosmic microwave background, apparently showing the initials "SH" near the centre of this image (NASA image)

the chances of finding other apparently improbable patterns may also be quite high [18].

The same point was made, in a slightly different way, in another of the arXiv April Fool spoofs—"Pi in the Sky" (2016), once again authored by Douglas Scott and the anagrammatic Ali Frolop. Quoting from the paper's abstract:

Deviations of the observed cosmic microwave background (CMB) from the standard model, known as "anomalies", are obviously highly significant and deserve to be pursued more aggressively in order to discover the physical phenomena underlying them. Through intensive investigation we have discovered that there are equally surprising features in the digits of the number π, and moreover there is a remarkable correspondence between each type of peculiarity in the digits of π and the anomalies in the CMB. Putting aside the unreasonable possibility that these are just the sort of flukes that appear when one looks hard enough, the only conceivable conclusion is that, however the CMB anomalies were created, a similar process imprinted patterns in the digits of π [19].

Like all the arXiv spoofs, this one has all the appearances of a serious scientific paper. It's a particularly impressive example, in fact, using no fewer than 14 meticulously produced charts and diagrams to get its point across. It would, however, be uncharitable to suggest that Professor Scott has a little too much spare time on his hands.

That number π (the Greek letter pi) is, of course, the ratio of the circumference of a circle to its diameter. It's a famously "irrational" number, in the sense that it has an infinite number of essentially random, non-recurring digits after the decimal point. It's famous for another reason too—it's one of the fundamental constants of mathematics. Its value never changes—or does it?

Another of the arXiv spoofs, produced by Robert Scherrer in 2009, examines the "Time Variation of a Fundamental Dimensionless Constant"—the constant in question being pi. Scherrer argues (with tongue firmly in cheek) that pi has varied over time, based on the indisputable fact that its recorded numerical value has changed in the course of the last 4000 years as calculation methods have become more sophisticated [20].

This notion is closely related to a more widely known April Fool's joke. It took the form of a spoof press release, posted on an internet newsgroup on 1 April 1998, claiming that Alabama had passed a law stating that henceforth pi was exactly three. This was actually intended as another anti-creationism parody, in the same spirit as *Scientific American*'s "Okay, We Give Up" editorial mentioned earlier. Here is what the Snopes fact-checking site has to say about "Alabama's Slice of Pi":

> Written by Mark Boslough as an April Fool's parody on legislative and school board attacks on evolution in New Mexico, the author took real statements from New Mexican legislators and school board members supporting creationism and recast them into a fictional account detailing how Alabama legislators had passed a law calling for the value of pi to be set to the "Biblical value" of 3.0 [21].

And here is an excerpt from the hoax press release itself:

> The Alabama state legislature narrowly passed a law yesterday redefining pi, a mathematical constant used in the aerospace industry. The bill to change the value of pi to exactly three was introduced without fanfare by Leonard Lee Lawson (Republican, Crossville), and rapidly gained support after a letter-writing campaign by members of the Solomon Society, a traditional values group. Governor Guy Hunt says he will sign it into law on Wednesday...

> Professor Kim Johanson, a mathematician from University of Alabama, said that pi is a universal constant, and cannot arbitrarily be changed by lawmakers. Johanson explained that pi is an irrational number, which means that it has an infinite number of digits after the decimal point and can never be known exactly. Nevertheless, she said, pi is precisely defined by mathematics to be "3.14159, plus as many more digits as you have time to calculate".

> "I think that it is the mathematicians that are being irrational, and it is time for them to admit it," said Lawson. "The Bible very clearly says in *I Kings* 7:23 that the altar font of Solomon's Temple was ten cubits across and thirty cubits in diameter, and that it was round in compass."

By a curious coincidence, this real-world spoof had a science-fictional precursor in the form of Robert A. Heinlein's 1961 novel *Stranger in a Strange Land*. When recounting the events that occur after the protagonist, Mars-born Valentine Michael Smith, returns to Earth and becomes a celebrity, Heinlein includes the following snippet:

> In the Tennessee legislature a bill was introduced to make pi equal to three; it was reported out by the committee on public education and morals, passed without objection by the lower house and died in the upper house [22].

Returning to Zaslavsky's list of arXiv spoofs: as extensive as it is, it isn't completely comprehensive, and a few others are worth a mention. From 2015, for example, there's S. E. Kuhn's serious-sounding "Observation of a New Type of Super-Symmetry". It turns out, however, that the symmetry in question has more to do with geography than physics or mathematics. The abstract begins as follows:

> We report the discovery of an unexpected symmetry that correlates the spin of all elementary particles (integer versus half-integer) with the geographic location of their initial discovery [23].

This is yet another twist on the "spurious correlation" theme, with the correlation in this case being spatial rather than temporal. It's a perfectly real correlation too, based on genuine historical data, but there's no profound significance to it. It has to do with the "Standard Model" of particle physics, which as shown in Fig. 3 involves 17 fundamental particles. Five of these are "bosons", which have an integer value of spin, and 12 are "fermions", with half-integer spin.

It just so happens that all five bosons, from the photon to the Higgs boson, were discovered in continental Europe, while all 12 fermions, from the electron to the top and bottom quarks, were first detected in the English-speaking nations of the United Kingdom or the United States. That's an interesting piece of trivia—and it's the sum total of the "New Type of Super-Symmetry" reported in Kuhn's paper.

Although several of the arXiv spoofs are satirical in nature, the target of the satire is usually some fairly specialized aspect of physics itself. One notable exception to this rule dates from April 2017. Tom Banks's spoof paper "Schrodinger's Cat and World History" centres on a topical news event of the time that everyone knew about. Against all the predictions of media pundits and polling agencies, the businessman and TV celebrity Donald Trump was elected President of the United States. It's this "low probability event" that Banks addresses in his paper. As he explains in the abstract:

> I propose that much recent history can be explained by hypothesizing that sometime during the last quarter of 2016, the history of the world underwent a macroscopic quantum tunnelling event, creating, according to the many-worlds interpretation, a new branch of the multiverse.

Just what he's getting at becomes clearer later in paper:

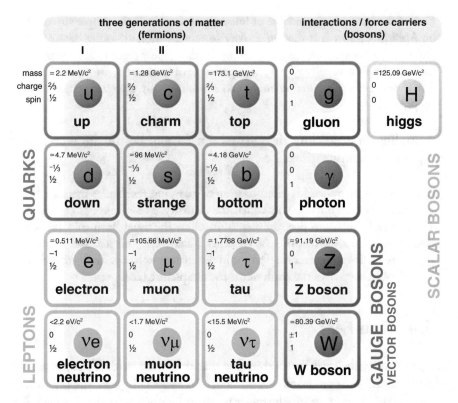

Fig. 3 The 17 fundamental particles of the "standard model". All the fermions were first detected in Britain or North America, and all the bosons in continental Europe (Wikimedia user Miss MJ, CC-BY-3.0)

According to the many-worlds interpretation, the different universes accessed by vacuum tunnelling are equally real, and can only communicate with each other by highly improbable tunnelling events… The new universe is similar in many respects to our own, except for a few improbable events which change the course of history… In short, the hypothesis of this paper is that the improbable result of the election for the Presidency of the United States was the result of a low amplitude tunnelling event in the wave function of the universe. This simple hypothesis explains at a stroke why the highly scientific statistical methodologies failed to predict this occurrence [24].

This is just the latest manifestation of something SF writers and pseudoscientists have known for a long time. It's possible to justify any theory, no matter how far-fetched, simply by invoking quantum physics. In an earlier arXiv spoof, from 2014, George Svetlichny applied this same principle to April Fool's day itself. In "The April First Phenomenon" he wrote:

One of the least understandable of human phenomena is the propensity to fib on April first. How is it that an activity so reprehensible on other days of the year is so readily accepted on this one singular day? Many theories have been put forth about this, usually of a sociological type. However … we have to seek deeper causes for the phenomenon, and obviously only quantum physics can supply this.

According to Svetlichny the answer, once again, lies in the ever-helpful many-worlds interpretation:

In the Everett many-worlds interpretation of quantum mechanics there are the so called maverick universes in which the ordinary laws of physics can break down because quantum probabilities don't follow the usual Born rule. It must surely be that on each April first we enter a maverick universe and so what appears to be fibs are in fact solid truths in the current universe. This settles that. April Fools' Day proves the truth of the Everett picture [25].

A Joke Too Far?

The problem with April Fool's spoofs is that they only work if people get the joke (sometimes after a moment's thought). In a specialist area like physics, or any of the other sciences, the humour may not be obvious to a non-specialist. For example, on the 1st of April 1923, the German newspaper *Deutsche Allgemeine Zeitung* ran a physics-based spoof which was picked up, a couple of days later, as a "real" news report by the *New York Times*. Here is what the *Museum of Hoaxes* website says about the case:

The *Deutsche Allgemeine Zeitung* reported that a Russian scientist, Professor Figu Posakoff, had discovered a method of "harnessing the latent energy of the atmosphere", the energy displayed in thunderstorms and other atmospheric catastrophes. Harnessing this energy would allow the Soviets to hurl objects "of any weight almost unlimited distances"… The Soviets were said to have promised to use this discovery only for peaceful purposes. However the *Allgemeine Zeitung* noted that it would certainly give the nation a powerful advantage in warfare. The *New York Times* ran the story on its front page on April 3, having failed to realize that it was a joke [26].

The result was a prominent headline in the *New York Times* announcing that: "Russian Claims Harnessing of Air Energy; Soviet Holds It is Greatest Discovery" (Fig. 4).

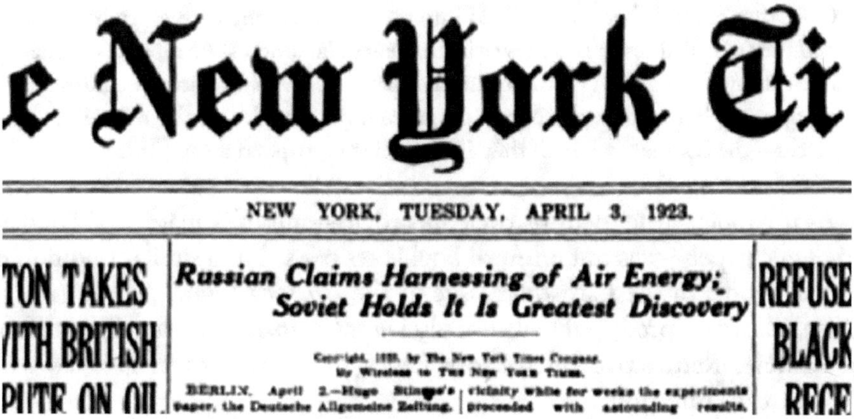

Fig. 4 The headline in the *New York Times* from 3 April 1923, repeating a physics-based April Fool's joke as if it were real news (public domain image)

To most people, seeing someone fall for an April Fool is very funny. To others, however, it can be worrying—at least when the subject is a scientific one. In 2015, the journal *Science and Engineering Ethics* published a paper by Maryam Ronagh and Lawrence Souder called "The Ethics of Ironic Science in Its Search for Spoof". Here is part of the abstract:

> The goal of most scientific research published in peer-review journals is to discover and report the truth. However, the research record includes tongue-in-cheek papers written in the conventional form and style of a research paper. Although these papers were intended to be taken ironically, bibliographic database searches show that many have been subsequently cited as valid research, some in prestigious journals… Some citing authors interpret the research as valid and accept it, some contradict or reject it, and some acknowledge its ironic nature. We conclude that publishing ironic science in a research journal can lead to the same troubles posed by retracted research [27].

By "retracted research", the authors refer to work that was published in good faith but later withdrawn by the authors after the results were discovered to be erroneous—a situation that occasionally arises in all fields of science. The problem is, nothing can be completely erased once it has been published—it will still be present in hard-copy libraries and online databases.

Writing in *The Atlantic* in the wake of Ronagh and Souder's paper, Rose Eveleth echoed their point that scientific April Fools can be a bad thing: "once the laughs have worn off, spoof papers can actually do damage to science". She went on:

Once they're published, they're filed into the archives along with everything else, and they're called up in searches as if they're regular studies. As more researchers move away from reading journals in their paper form—in which an editorial or an opening letter from the editor might remind a reader of the nature of the pieces—the context in which they initially live is stripped away [28].

As it happens, the focus of Eveleth's article—and of Ronagh and Souder's academic paper—was not on April Fool jokes per se but a similar tradition of annual spoofs in the Christmas issue of the *British Medical Journal* (BMJ). One such spoof, produced by Leonardo Leibovici in 2001, was called "Effects of Remote, Retroactive Intercessory Prayer on Outcomes in Patients with Bloodstream Infection".

That title in itself should be a giveaway. It's not just that mainstream medical researchers are unlikely to take an interest in the efficacy of prayer, but the addition of the word "retroactive" implies that the prayers are being offered backwards in time. As ridiculous as that scenario is, it's treated with a straight face by Leibovici, who asserts that "we cannot assume a priori that time is linear, as we perceive it, or that God is limited by a linear time, as we are".

Nevertheless—perhaps because of the paper's realistically academic style—not everyone realized it was a spoof. To quote from Eveleth's article again:

Leibovici's paper was one of many of BMJ's Christmas spoofs, appearing in the journal alongside other joke articles. But eight years later the paper was cited, unironically, in a review paper from a well-respected organization.

She was sufficiently intrigued to carry out a small survey of her own:

I searched for a few past Christmas issue studies to see where they've showed up since. One joke study from 2007 on the energy expenditure of adolescents playing video games has been cited about 400 times since then, according to a Google Scholar estimate…

A study called "sex, aggression, and humour: responses to unicycling" was cited in 2012 as evidence for "the evolution of humour from male aggression" and appears in a book called *The Male Brain*. In fact, that unicycle study wasn't just cited by other scientists, it was picked up by the BBC for a story with the headline "Humour comes from testosterone" [28].

Another negative opinion of spoofs comes from science writer Lee Billings, who wrote a piece called "Against April Fools in Science Journalism" for the *Scientific American* website in 2015. He admits he has a personal grudge against the genre, after falling, as a young school student in 1996, for the "Bigon" spoof in *Discover* magazine that was mentioned earlier in this chapter.

To any adult reader, that piece is full of obvious jokes. The discoverer's name is Manqué—French for "failed"—and his "main research consists of building better vacuum tubes to replace microchips" [7]. Most adults in the 1990s, recalling the huge, clunky vacuum tubes of the pre-transistor era, would have found that hilarious. On the other hand, to a youngster brought up in a world where microchips are commonplace—and "vacuum tube" might sound like an exotic new kind of technology—it's a different matter.

Having been taken in himself, Billings has sympathy for others in the same situation:

> There is a long, illustrious history of highly regarded journalists and publica-tions pranking readers with goofy stories on April Fool's Day…. But perhaps it's past time for reputable science publications to abandon the practice—or at least to quietly discourage it. What seems like harmless fun among journalists and their more-savvy readers may have negative unintended consequences outside those knowledgeable inner circles. Recent polling data shows that public trust in science and scientists is not exactly stellar, and ongoing controversies over topics such as climate change, evolution, vaccines and genetically-modified organisms illustrate how easily insidious forces can manipulate the media to promote unscientific agendas. It's unclear—to me at least—how tongue-in-cheek articles designed to betray a reader's trust could possibly do anything but exacerbate these serious problems [29].

That's a valid opinion, of course—but it's essentially that of a killjoy, and hopefully not too many people share it. April Fools and similar spoofs are meant to be funny, and (unless the perpetrator is extremely inept) that should be obvious enough to anyone who reads them with sufficient attention.

There is, however, another kind of "fake physics", which really is designed to take in the intended victim—even if the rest of the world can see at a glance that it's an obvious joke. That's the subject of the next chapter, "Making a Point".

References

1. S. Lee, S. Ditko, I come from the Black Void. reprinted in Creatures on the Loose **#27** (January 1974)
2. BBC *On This Day*, 1957: BBC fools the nation. http://news.bbc.co.uk/onthis-day/hi/dates/stories/april/1/newsid_2819000/2819261.stm

3. Editorial criteria and processes. Nature. https://www.nature.com/nature/for-authors/editorial-criteria-and-processes

4. A.J. Hamilton, R.M. May, E.K. Waters, Here be dragons. Nature **520**, 42–43 (April 2015)

5. J. Rennie, Okay, we give up. Scientific American (April 2005). https://www.scientificamerican.com/article/okay-we-give-up/

6. P. Yam, Unscientific Unamerican. Scientific American (April 2014). https://blogs.scientificamerican.com/observations/unscientific-unamerican-and-other-april-fools-jokes-in-sa-history/

7. T. Folger, The Bigon. Discover (April 1996). http://discovermagazine.com/1996/apr/thebigon756

8. A. Boese, The black-hole diode. Museum of Hoaxes. http://hoaxes.org/af_database/permalink/the_black_hole_diode

9. Apollo Bacteria Spur Lunar Erosion. Nature Website (1 April 2005). http://www.bioedonline.org/news/nature-news/apollo-bacteria-spur-lunar-erosion/

10. C. O'Luanaigh, CERN researchers confirm existence of the force. CERN Website (1 April 2015). https://home.cern/news/news/cern/cern-researchers-confirm-existence-force

11. Wikipedia article on "April Fools" day request for comments. https://en.wikipedia.org/wiki/April_Fools%27_Day_Request_for_Comments

12. D. Waitzman, A standard for the transmission of IP datagrams on Avian Carriers. Network Working Group RFC 1149 (April 1990). http://www.faqs.org/rfcs/rfc1149.html

13. R. Hinden, Design considerations for faster-than-light communication. Check Point Software RFC 6921 (1 April 2013). https://tools.ietf.org/html/rfc6921

14. D. Zaslavsky, Joke papers. arXiv. https://www.ellipsix.net/arxiv-joke-papers.html

15. S.R. Kane, F. Selsis, A Necro-biological explanation for the Fermi Paradox. arXiv (31 March 2014). https://arxiv.org/abs/1403.8146

16. D. Scott, A farewell to falsifiability. arXiv (1 April 2015). https://arxiv.org/abs/1504.00108

17. J. Zuntz, et al., Non-standard morphological relic patterns in the cosmic microwave background. arXiv (31 March 2011). https://arxiv.org/abs/1103.6262

18. R. Fisher, R. Courtland, Found: Hawking's initials written into the universe. New Scientist, (7 February 2010). https://www.newscientist.com/article/dn18489-found-hawkings-initials-written-into-the-universe/

19. D. Scott, Pi in the sky. arXiv (31 March 2016). https://arxiv.org/abs/1603.09703

20. R.J. Scherrer, Time variation of a fundamental dimensionless constant. arXiv (30 March 2009). https://arxiv.org/abs/0903.5321

21. Snopes Fact Check, Alabama's slice of Pi. https://www.snopes.com/fact-check/alabamas-slice-of-pi/

22. R.A. Heinlein, *Stranger in a Strange Land* (New English Library, London, 1978), p. 292

23. S.E. Kuhn, Observation of a new type of super-symmetry. arXiv (31 March 2015). https://arxiv.org/abs/1503.09109

24. Tom Banks, Schrodinger's cat and world history. arXiv (30 March 2017). https://arxiv.org/abs/1703.10470

25. G. Svetlichny, The April first phenomenon. arXiv (28 March 2014). https://arxiv.org/abs/1403.8010

26. A. Boese, Atmospheric energy harnessed. Museum of Hoaxes. http://hoaxes.org/af_database/permalink/atmospheric_energy_harnessed

27. M. Ronagh, L. Souder, The ethics of ironic science in its search for spoof. Sci. Eng. Ethics **21**, 1537–1549 (2015)

28. R. Eveleth, The ethics of sarcastic science. The Atlantic (December 2014). https://www.theatlantic.com/technology/archive/2014/12/the-ethics-of-sarcastic-science/383988/

29. L. Billings, Against April fools in science journalism. Scientific American Website (1 April 2015). https://blogs.scientificamerican.com/observations/against-april-fools-in-science-journalism/

Making a Point

Abstract All the varieties of "fake physics" discussed so far, from sci-fi to April Fool jokes, were designed purely for entertainment purposes. Entertainment plays an important role in this chapter too—but the spoofs and hoaxes described here all had another, more serious purpose behind them. To start with, we look at a few of the much-publicized "sting operations" that have been used to trick journals and conferences with very low editorial standards. Next comes the famous Sokal hoax—and others like it—where the target shifts from lazy editors to politically partisan ones. Finally, we consider the ways in which physics has been affected by the current fashion for "fake news".

Scientific Writing, the Lazy Way

The computer program SCIgen has already been mentioned, in the chapter on "The Art of Technobabble". Its purpose is to generate random but superficially convincing-looking academic papers in, appropriately enough, the field of computer science. It was originally written simply as an amusing in-joke for computer programmers, but almost immediately it found a more serious use. As SCIgen's developers, Jeremy Stribling, Dan Aguayo and Max Krohn, say on their website:

> One useful purpose for such a program is to auto-generate submissions to conferences that you suspect might have very low submission standards... Using SCIgen to generate submissions for conferences like this gives us pleasure to no end [1].

Stribling et al. first put SCIgen to work in 2005, when they used it to create a randomly-generated paper called "Rooter: A Methodology for the Typical

© Springer Nature Switzerland AG 2019
A. May, *Fake Physics: Spoofs, Hoaxes and Fictitious Science*, Science and Fiction,
https://doi.org/10.1007/978-3-030-13314-6_6

Unification of Access Points and Redundancy". They submitted this to an upcoming conference that had a high attendance fee but (they suspected) low editorial standards. The ploy was successful, and widely reported in the mainstream media—such as this account from BBC News:

> A collection of computer-generated gibberish in the form of an academic paper has been accepted at a scientific conference, to the delight of hoaxers. Three US boffins built a program designed to create research papers with random text, charts and diagrams… One of the hoaxers said the fake paper was designed to expose the lack of standards at academic gatherings… It was accepted for the World Multiconference on Systemics, Cybernetics and Informatics, due to be held in the city of Orlando in July [2].

Because SCIgen is free to use, many other pranksters have followed suit, and over the years numerous papers created by it have been accepted for conferences around the world. In 2014, *Nature* reported:

> Over the past two years, computer scientist Cyril Labbé of Joseph Fourier University in Grenoble, France, has catalogued computer-generated papers that made it into more than 30 published conference proceedings between 2008 and 2013. Sixteen appeared in publications by Springer, which is headquartered in Heidelberg, Germany, and more than 100 were published by the Institute of Electrical and Electronic Engineers (IEEE), based in New York. Both publishers, which were privately informed by Labbé, say that they are now removing the papers…

> Labbé developed a way to automatically detect manuscripts composed by a piece of software called SCIgen, which randomly combines strings of words to produce fake computer-science papers… Labbé has emailed editors and authors named in many of the papers and related conferences but received scant replies; one editor said that he did not work as a programme chair at a particular conference, even though he was named as doing so, and another author claimed his paper was submitted on purpose to test out a conference, but did not respond on follow-up [3].

At first sight it may seem shocking that high-profile organizations like Springer and the IEEE were caught out in this way so many times, but their role was simply that of publisher. Unlike the academic journals produced by these organizations, the contents and quality assurance of conference proceedings lies with the conference organizers and their own peer-reviewing procedures. Clearly in these cases too little effort went into the latter.

There are many reasons why this might be the case, but perhaps the most obvious is that cutting corners saves money. Not all conferences that are run for profit are money-making scams—not even all the ones that have been taken in by SCIgen—but some of them are. A case that hit the headlines in October 2016 is notable because it relates to the main subject of this book, physics. It concerns another paper than was accepted for publication despite being completely nonsensical—and it wasn't even written by a physicist, as the *Guardian* reported at the time:

Christoph Bartneck, an associate professor at the Human Interface Technology laboratory at the University of Canterbury in New Zealand, received an email inviting him to submit a paper to the International Conference on Atomic and Nuclear Physics in the US in November. "Since I have practically no knowledge of nuclear physics I resorted to iOS autocomplete function to help me writing the paper," he wrote in a blog post on Thursday. "I started a sentence with 'atomic' or 'nuclear' and then randomly hit the autocomplete suggestions" [4].

The paper's title, also created via autocomplete, was "Atomic Energy Will Have Been Made Available to a Single Source". Here is a small excerpt from it, as quoted by the *Guardian*:

The atoms of a better universe will have the right for the same as you are the way we shall have to be a great place for a great time to enjoy the day you are a wonderful person to your great time to take the fun and take a great time and enjoy the great day you will be a wonderful time for your parents and kids.

Now that's real nonsense. SCIgen's nonsense does at least look fairly convincing to a non-specialist, but autocomplete's effort is in a different league; even a child could recognize it as nonsense. Nevertheless, to continue from the *Guardian* article:

The nonsensical paper was accepted only three hours later, in an email asking Bartneck to confirm his slot for the "oral presentation" at the international conference… The acceptance letter referred him to register for the conference at a cost of US $1099 [4].

Returning for a moment to SCIgen—here's a slightly different case, as recounted in *New Scientist* in 2009:

Philip Davis, a graduate student at Cornell University in Ithaca, New York … got a nonsensical computer-generated paper accepted for publication in a peer-reviewed journal. Earlier this year, Davis started receiving unsolicited emails from Bentham Science Publishers, which publishes more than 200 "open-access" journals… Davis teamed up with Kent Anderson, a member of the publishing team at *The New England Journal of Medicine*, to put Bentham's editorial standards to the test. The pair turned to SCIgen, a program that generates nonsensical computer science papers, and submitted the resulting paper to *The Open Information Science Journal*, published by Bentham…

Davis and Anderson, writing under the noms de plume David Phillips and Andrew Kent, also dropped a hefty hint of the hoax by giving their institutional affiliation as the Centre for Research in Applied Phrenology, or CRAP. Yet four months after the article was submitted, "David Phillips" received an email from Sana Mokarram, Bentham's assistant manager of publication: "This is to inform you that your submitted article has been accepted for publication after peer-reviewing process in TOISCIJ"…. The publication fee was $800, to be sent to a PO Box in the United Arab Emirates. Having made his point, Davis withdrew the paper [5].

This is as amusing as the previous examples, but there's a subtle difference—the target was not a conference but a journal. In the former case, the financial motive behind "low editorial standards" is obvious: the more people that can be persuaded to attend a conference, the more money it will make. As it happens, there's a similar—but less obvious—equation in the case of certain types of journal. That's the subject of the next section.

Predatory Journals

As we saw in the chapter "The Art of Technobabble", the publication of results is of critical importance in science. In the traditional model, as with most other types of publishing, authors aren't charged for their contributions. Instead, journals make their money through subscriptions and single-copy sales to readers. However, the twenty-first century has seen the rise of an alternative in the form of "open access" journals. Here is what a traditional journal, *Nature*, said on the subject in 2013:

In the conventional subscription-based model, journals bring in revenue largely through selling print or web subscriptions and keeping most online content locked behind a paywall. But in the most popular model of open access, publishers charge an upfront "author fee" to cover costs—and to turn a profit, in the case of commercial publishers—then make the papers freely available online, immediately on publication [6].

The open access sector has seen a huge boom since the start of the twenty-first century, dramatically demonstrated in statistics collected by Mikael Laakso and Bo-Christer Björk [7]. Their data on the growth of open access publishing in the physical sciences is shown in Fig. 1.

On the face of it, open access is a good thing. It gives scientists a greater range of outlets in which to publish their results, and—because these are freely available to readers—they are more likely to be read. Unfortunately, as with any successful business model, there will always be a few unscrupulous practitioners who abuse it to their financial advantage.

This phenomenon was first noticed by academic librarian Jeffrey Beall, who coined the term "predatory journals" to describe it. Quoting from *Nature* again:

> Open-access publishers often collect fees from authors to pay for peer review, editing and website maintenance. Beall asserts that the goal of predatory open-access publishers is to exploit this model by charging the fee without providing all the expected publishing services [6].

It's important to state that this isn't true of most open access publishers, who give good value for money. If that wasn't the case, the open access sector would have collapsed long ago. Instead, it's thriving—and that provides an irresistible temptation to certain people to masquerade as the real thing, charging authors for non-existent peer review and editing services.

The subject of predatory journals is an unsavoury one, because it takes money from well-meaning academics in an unethical way, and devalues the whole notion of open access in the process. From the point of view of this book, however, predatory publishing has had one positive benefit. It's given rise to a whole new genre of scientific spoofs.

The idea is simple enough. If a predatory journal charges authors an extortionate sum for printing articles, and if it has no intention of offering any

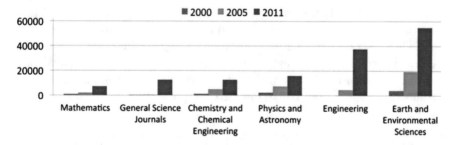

Fig. 1 The number of articles published in open access journals in various physical science disciplines, in the years 2000, 2005 and 2011 (after reference [7], CC-BY-2.0)

kind of peer review or copy editing services, then it will pretty much accept anything—no matter how outrageous or comical.

The practice of catching out a predatory journal with a fake submission has become known as a "sting", after the term used in law enforcement for a deceptive operation designed to catch a person in the act of committing a crime. Overcharging, which is all predatory publishing amounts to, is hardly a crime—but it's profoundly unethical, and ethics are very important in academia.

Despite the seriousness of the context, many stings are highly amusing. Take the one perpetrated by an anonymous scientist calling himself "Neuroskeptic", as he explained on *Discover* magazine's website in July 2017:

> A number of so-called scientific journals have accepted a *Star Wars*-themed spoof paper. The manuscript is an absurd mess of factual errors, plagiarism and movie quotes. I know because I wrote it. Inspired by previous publishing "stings", I wanted to test whether predatory journals would publish an obviously absurd paper. So I created a spoof manuscript about "midichlorians"—the fictional entities which live inside cells and give Jedi their powers in *Star Wars*. I filled it with other references to the galaxy far, far away, and submitted it to nine journals under the names of Dr Lucas McGeorge and Dr Annette Kin [8].

Those names, of course, recall *Star Wars* creator George Lucas and the villainous Darth Vader's alter ego, Anakin Skywalker. The fictitious word "midichlorians" is loosely based on a real scientific term, mitochondria, referring to a microscopic component of living cells. That's the term in the title of Neuroskeptic's spoof, "Mitochondria: Structure, Function and Clinical Relevance"—but you don't have to read far into the abstract before those midichlorians turn up:

> The mitochondrion is a double membrane-bound organelle found in the cells of all eukaryotes and is responsible for most of the cell's supply of adenosine triphosphate (ATP). As the central powerhouse of the cell‖, mitochondria (also referred to as midichlorians) serve a vital function and they have been implicated in numerous human diseases, including midichlorial disorders, heart disease and circulatory failure, and autism. In this paper, the structure and function of the midichlorian is reviewed with a view to understanding how the pathophysiology of midichlorial disorders can point the way towards translational treatments [9].

Another anonymous author, going by the online handle BioTrekkie, produced a similar spoof called "Rapid Genetic and Developmental Morphological

Change Following Extreme Celerity". As you might guess from the pseudonym, this one concerns *Star Trek* rather than *Star Wars*. It too was accepted for publication in spite of its ludicrous subject-matter—and the fact that its author list included the names of Thomas Paris, Harry Kim, B'Elanna Torres, Kes Ocampa and Kathryn Janeway. Those are all characters from the *Star Trek: Voyager* TV series.

According to a news item that appeared on the Space.com website in February 2018:

In the *Star Trek* universe, the fantastic speed of warp 10 has remained annoyingly out of reach. However, a recent paper in an open-access journal describes an experiment that attempted to break that boundary. The fact that the "experiment" described in the paper wasn't conducted in a real-world laboratory, but in an episode of the sci-fi TV series *Star Trek: Voyager*, reveals just how easy it is to publish fake science in some so-called "predatory journals"... BioTrekkie's research follows *Voyager* episode 32, "Threshold", where Lieutenant Thomas Paris rigs an experimental shuttle to cross the warp 10 boundary... The paper describes the plot of the episode as research. The paper uses the word "celerity", which means extreme speed, in its title, but further down BioTrekkie parenthetically describes the theoretical maximum celerity as "warp 10". Despite this and other fictions, including those involving the author and affiliation, the paper was published as legitimate research [10].

Here's a short extract from the paper itself:

We employed a replicated design wherein the two human subjects were exposed to the theoretical maximum celerity (warp 10) and examined... Physical responses to the celerity became apparent in later observations. Spontaneous exfoliation of skin cells commenced, and a comparably thick intact layer of new skin cells formed within 96 hours. Internal morphological differences were noted via MRI and *ImageJ* analysis, with measurement of heart number increasing two-fold [9].

Compare that with the relevant episode summary on the *Star Trek* fan site *Memory Alpha*:

Paris's entire body is mutating and his lungs are no longer processing oxygen. The doctor replaces the atmosphere in the room with 80% nitrogen and 20% acidichloride. Paris can now breathe but a bigger problem has also developed; he is suffering from cellular degradation and is consequently dying... The doctor scans him and finds something peculiar; Paris now has two hearts [11].

The same web page also notes that:

> This episode was panned by critics, frequently being voted as the worst ever episode of *Star Trek: Voyager* and even the worst episode of *Star Trek* in general.

To date, the total number of "stings" of this type, most of them in fields other than physics, is enormous—so it's surprising that journals are still falling for them on a regular basis. Another sci-fi-related one appeared in October 2018—this time relating to the animated series *Rick and Morty*. Less familiar to most people than *Star Wars* and *Star Trek*, this show began life as a pastiche of *Back to the Future*, with the title characters loosely based on Doc Brown and Marty McFly (see Fig. 2).

Appearing, among other places, in volume 13 of *IOSR Journal of Pharmacy and Biological Sciences*, the *Rick and Morty* paper—supposedly authored by Beth Smith et al.—was called "Newer Tools to Fight Intergalactic Parasites and their Transmissibility in Zyrgion Simulation". It's written in a nonsensical style reminiscent of the famous "Turboencabulator" piece, which was described in the chapter "Spoofs in Science Journals", as you can see from this brief sample:

Fig. 2 Sci-fi cartoon characters Rick and Morty, who inspired a "sting" type spoof paper in 2018 (Wikimedia user Danielobich 23, CC-BY-SA-4.0)

Briefly, a dinglebop was smoothened by the help of schleem. The obtained product was then subjected to ultrasonication, and repurposed for later batches. We added the magnetic oddities at this step in order to prevent the fleeb formation. This was called our oil phase and this was added to an aqueous phase under constant schwitinization until a homogenous mixture was obtained [9].

Here is what an article on the *Vice* website has to say about it:

Beth Smith, the Zyrgion simulation, and intergalactic … parasites are all references to *Rick and Morty*. The paper is an obvious troll, but that didn't stop three scientific journals—*ARC Journal of Pharmaceutical Sciences*, *IOSR Journal of Pharmacy and Biological Sciences* and *Clinical Biotechnology and Microbiology* from publishing the paper without a second glance.

Unlike the *Star Wars* and *Star Trek* stings, the author of this one was happy to be identified. It was Farooq Ali Khan, an undergraduate college professor and PhD student in Hyderabad, India. *Vice* quotes him as follows:

The fake science, fake news epidemic is getting worse by every day, and I really wanted to do something about it. There's a lot of money involved in it and these people are getting more powerful, and several mediocre science papers are being published, which is a severe threat to science and academic research [12].

In all the foregoing examples, the target journals were motivated purely by profit. They fell for the spoofs either because they habitually accept papers without reading them, or because they read the papers, got the joke—and then published them anyway, because there was no financial downside in doing so.

At the other extreme are journals where the editors are motivated not by money but by firmly held principles. That makes them vulnerable to a completely different type of spoof—one that carefully panders to those principles while appearing ludicrous to the world at large. That's the subject of the next section.

Science Wars

Earlier in this book, the chapter on "The Relativity of Wrong" gave a brief overview of the philosophical background to science. Its scope is limited to the physical world, and to phenomena that are amenable to experimental investigation in the spirit of "falsifiability". A non-scientist could criticize science on either of these grounds, because it ignores a vast range of human experience in the fields of aesthetics, religion and politics. That's a perfectly valid criticism—in fact it's a limitation of science that many scientists recognize.

On the other hand, there's a much more fundamental assumption underlying science that no scientist would ever question—or even realize that it could be questioned. This is the idea that the focus of scientific study, the physical world, has an objective existence that is independent of human consciousness. As "obvious" as this is to people with a scientific upbringing, it's far from obvious to others. It led to a bizarre dispute in American academia in the 1990s, which became known by the dramatic name of "science wars".

The conflict is summarized by Michael Lynch, a professor of Science and Technology Studies at Cornell University, in the following way:

> As usually portrayed, the science wars involve a conflict between two opposing camps: natural scientists and sociologists. The sociologists are identified with the far left of the political spectrum, while the scientists … are associated with the right. The scientists are said to believe in nature, truth and reality, while the sociologists are said to believe that representations of nature are arbitrary, scientific laws are ideological, and reality is a myth [13].

For the scientist's viewpoint, Lynch quotes a *Physics World* editorial from 1997:

> Many scientists feel uneasy about various ideas from sociology, notably the suggestion that the laws of nature as we know them are social constructs—essentially laws that scientists have agreed between themselves—and do not have any fundamental significance.

In over-simplified terms, one might say the establishment scientists of the 1990s were grown-up versions of the bespectacled, geeky teenagers who read superhero comics in the 1960s and devoured the works of Robert Heinlein. So what happened to their swinging, pot-smoking, hippie contemporaries? They went the other way, suggests Lynch:

> The sociology side is generally identified with radical leftist politics, and some writers argue that student radicals from the 1960s have grown up to become professors and academic administrators who dominate particular departments and colleges [13].

The reality, of course, was much more nuanced than this. In the context of the present book, however, it's worth emphasizing the cartoonish aspects because of one particularly comical skirmish in the science wars. This was the notorious Sokal hoax, which took place in 1996. It was prompted by a special "science wars" issue of a distinctly left-of-centre journal called *Social Text*. To quote philosopher of science Ziauddin Sardar:

> It was Paul Gross and Norman Levitt's *Higher Superstition: the Academic Left and its Quarrels with Science*, published a year earlier, that more than anything else motivated the *Social Text* issue on "science wars". Biologist Gross and mathematician Levitt declared that the academic left ... was intrinsically anti-science. This anti-science hostility is based not just on the academic left's dislike of the uses to which science and technology are put by political and economic forces—such as military hardware, surveillance, industrial pollution and destruction of the environment. Even scientists regret these abuses of science and technology. The hostility "extends to the social structures through which science is institutionalized, and to a mentality that is taken, rightly or wrongly, as characteristic of scientists. More surprisingly, there is open hostility toward the actual content of scientific knowledge and toward the assumption, which one might have supposed universal among educated people, that scientific knowledge is reasonably reliable and rests on a sound methodology." [14]

When the editors of *Social Text* announced their intention to produce a special issue as a counterblast to this accusation, physicist Alan Sokal at New York University decided to submit a paper of his own to it. It was to be a spoof—and it turned out to be the most famous spoof in the history of physics.

Sokal's paper dealt with the highly specialized area of quantum gravity, so it would have been easy for him to fill it with "in-jokes" designed to go over the heads of the editors of *Social Text*. So easy, in fact, that the joke wouldn't have been very funny even if the editors had fallen for it. Sokal, however, played much fairer than that, by producing a spoof that even a non-specialist ought to spot if they gave it more than a cursory glance.

Under those circumstances, it really would be a coup if the editors accepted Sokal's piece and printed it in their special "science wars" issue. Yet that's exactly what they did, as Sardar explains:

> A reasonably critical examination of Sokal's paper would easily have aroused the suspicions of the editors. The paper purports to argue that unifying the currently incompatible theories of quantum mechanics and general relativity would produce a postmodern, "liberatory" science. It contains some deliciously daft

assertions. For example, it suggests that pi, far from being a constant and universal, is actually relative to the position of an observer and is thus subject to "ineluctable historicity" [14].

The title of Sokal's paper was "Transgressing the Boundaries: Towards a Transformative Hermeneutics of Quantum Gravity". That word "hermeneutics" despite sounding like archetypal sci-fi technobabble, can actually be in the dictionary. The problem is, it has nothing whatsoever to do with the subject of the paper, referring instead to "the branch of knowledge that deals with interpretation, especially of the Bible or literary texts".

A malapropism of that kind is a much subtler form of humour than a made-up word would have been. On a broader scale, that's the modus operandi of the whole paper. Almost every name, reference, technical term and theory mentioned in it is perfectly real—but jumbled together in an irrational mishmash of concepts from physics, biology, psychology, the arts, philosophy and mysticism. Here is a small taster:

> More recently, Lacan's *topologie du sujet* has been applied fruitfully to cinema criticism and to the psychoanalysis of AIDS. In mathematical terms, Lacan is here pointing out that the first homology group of the sphere is trivial, while those of the other surfaces are profound; and this homology is linked with the connectedness or disconnectedness of the surface after one or more cuts. Furthermore, as Lacan suspected, there is an intimate connection between the external structure of the physical world and its inner psychological representation qua knot theory: this hypothesis has recently been confirmed by Witten's derivation of knot invariants (in particular the Jones polynomial) from three-dimensional Chern-Simons quantum field theory [15].

Unlike, say, Asimov's "Thiotimoline" spoof, all the people named here are real. Edward Witten is a theoretical physicist specializing in string theory, while Jacques Lacan was a post-Freudian psychological theorist (the fact that there's absolutely no meaningful connection between Witten's work and Lacan's is neither here nor there). When you get into it, Sokal's style has a peculiar, *sui generis*, appeal. Here is another sample:

> An exciting proposal has been taking shape over the past few years in the hands of an interdisciplinary collaboration of mathematicians, astrophysicists and biologists: this is the theory of the morphogenetic field. Since the mid-1980s evidence has been accumulating that this field, first conceptualized by developmental biologists, is in fact closely linked to the quantum gravitational field: (a) it pervades all space; (b) it interacts with all matter and energy, irrespective of

whether or not that matter/energy is magnetically charged; and, most significantly, (c) it is what is known mathematically as a "symmetric second-rank tensor" [15].

A "morphogenetic field" is a metaphysical concept popularized by the maverick biochemist and author Rupert Sheldrake. It embodies the idea that the final shape of an organism is present, in an immaterial, ghostly way, from the very start. As Sheldrake says on his website:

> The oak tree has a morphogenetic field containing an attractor, in this case the mature form of the oak, which draws the developing acorn towards it. It plays the same role as what Aristotle called entelechy, the attractor within the soul [16].

This suggestion isn't an intrinsically crazy one, and Aristotle's ideas dominated Western philosophy for over a thousand years. It's just that modern science has shown that Aristotle was wrong. These days, you can't mention morphogenetic fields in the same breath as gravitational fields unless you're making a joke—and that, of course, is exactly what Sokal is doing.

His paper concludes with a list of no fewer than 240 references—again, all of them perfectly real. A handful actually have some relevance to the subject of quantum gravity, such as Green, Schwarz and Witten's *Superstring Theory* (1987). Others, such as James Gleick's *Chaos: Making a New Science* (1987), also come from the world of mainstream science—but are totally irrelevant to the paper's ostensible subject. Still others take what might politely be described as a "spiritual" approach to science, such as Fritjof Capra's *The Tao of Physics* (1975) or Sheldrake's own *A New Science of Life* (1981).

Soon after the paper appeared in *Social Text*, Sokal confessed to the hoax in an article, "A Physicist Experiments With Cultural Studies", published in the magazine *Lingua Franca*. Here, Sokal explained his motivation:

> For some years I've been troubled by an apparent decline in the standards of intellectual rigour in certain precincts of the American academic humanities… So, to test the prevailing intellectual standards, I decided to try a modest (though admittedly uncontrolled) experiment: would a leading North American journal of cultural studies … publish an article liberally salted with nonsense if (a) it sounded good and (b) it flattered the editors' ideological preconceptions? [17]

Sokal's chief target was the professed belief among many sociologists that, as mentioned earlier, the "reality" studied by physicists is purely subjective.

That viewpoint was treated with deadpan seriousness in Sokal's spoof, but in the *Lingua Franca* piece he offers a much blunter response to it:

> In the second paragraph I declare, without the slightest evidence or argument, that "physical reality … is at bottom a social and linguistic construct". Not our theories of physical reality, mind you, but the reality itself. Fair enough: anyone who believes that the laws of physics are mere social conventions is invited to try transgressing those conventions from the windows of my apartment (I live on the 21st floor) [17].

More recently, another amusing spoof—in a similar vein to Sokal's—was reported in *The Spectator* magazine in June 2017. Twenty years on, academic fashions had changed, and the target this time was the "politically correct" view that if you're white and male, you're pretty much responsible for everything that's bad about the world—up to and including climate change. Here is a quote from the *Spectator* piece:

> Two US academics, Peter Boghossian and James Lindsay, recently … managed to get published in a social sciences journal a paper arguing that the penis is not in fact a male reproductive organ but merely a social construct and that, furthermore, penises are responsible for causing climate change. It ought to go without saying that their paper, "The Conceptual Penis as a Social Construct", was a spoof. Yet it was peer-reviewed by two supposed experts in gender studies, one of whom praised the way it captured "the issue of hypermasculinity through a multidimensional and nonlinear process", and the other of whom marked it "outstanding" in every applicable category.

As with its predecessor two decades earlier, the secret of the paper's success lay in the way it was carefully engineered to pander to the editors' political views. The *Spectator* goes on:

> Like Sokal's, the latest hoax was careful to observe all the fashionable left-wing pieties. "We suspected that gender studies is crippled academically by an overriding, almost religious belief that maleness is the root of all evil," the authors later observed [18].

In reality, of course, climate change is driven not by male reproductive organs but by rising carbon dioxide levels (see Fig. 3).

This brings us to an important point: the far left doesn't have a monopoly on stupidity. The problem lies with the word "far", not "left". Anyone who stretches an ideology to its extreme limits is likely to be more than a little

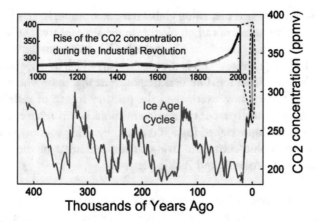

Fig. 3 Carbon dioxide levels have risen at an alarming rate since the start of the Industrial Revolution—but some people remain irrationally convinced that its due to any cause other than human activity (Wikimedia user Jklamo, CC-BY-SA-3.0)

irrational. An entertaining side effect of this is that such people often fail to recognize satire when they encounter it—which is precisely why *Social Text* fell for the Sokal hoax. At the other end of the political spectrum, it's just as easy to fool right-wingers who deny the reality of global warming.

This phenomenon has been mentioned already, in the chapter on "The Art of Technobabble", in the context of Michael Crichton's conspiracy-laden 2004 novel *State of Fear*. A few years later, in November 2007, the Reuters news agency carried the following report:

> A hoax scientific study pointing to ocean bacteria as the overwhelming cause of global warming fooled some sceptics on Thursday who doubt growing evidence that human activities are to blame. Laden with scientific jargon and published online in the previously unknown *Journal of Geoclimatic Studies* based in Japan, the report suggested the findings could be "the death of manmade global warming theory".

> Sceptics jumped on the report. A British scientist emailed the report to 2000 colleagues before spotting it was a spoof. Another from the US called it a "blockbuster"... But scientists knocked the report down. "The whole story is a hoax," Deliang Chen, professor of Meteorology at Gothenburg University in Sweden, told Reuters. He said two authors listed as from his university were unknown [19].

Despite its wholly spurious nature, the paper—"Carbon Dioxide Production by Benthic Bacteria"—is still listed on the ResearchGate website. Here is its abstract:

It is now well-established that rising global temperatures are largely the result of increasing concentrations of carbon dioxide in the atmosphere. The "consensus" position attributes the increase in atmospheric CO_2 to the combustion of fossil fuels by industrial processes. This is the mechanism which underpins the theory of manmade global warming. Our data demonstrate that those who subscribe to the consensus theory have overlooked the primary source of carbon dioxide emissions. While a small part of the rise in emissions is attributable to industrial activity, it is greatly outweighed (by >300 times) by rising volumes of CO_2 produced by saprotrophic eubacteria living in the sediments of the continental shelves fringing the Atlantic and Pacific oceans…

A series of natural algal blooms, beginning in the late nineteenth century, have caused mass mortality among the bacteria's major predators: brachiopod molluscs of the genus Tetrarhynchia. These periods of algal bloom, as the palaeontological record shows, have been occurring for over three million years, and are always accompanied by a major increase in carbon dioxide emissions, as a result of the multiplication of bacteria when predator pressure is reduced. They generally last for 150–200 years. If the current episode is consistent with this record, we should expect carbon dioxide emissions to peak between now and mid-century, then return to background levels. Our data suggest that current concerns about manmade global warming are unfounded [20].

It's a very clever spoof. Unlike the others mentioned so far in this chapter, it's not simply gobbledegook. It's still nonsense, but it's nonsense that hangs together in a logically consistent way. The central assertion—that oceanic bacteria are currently producing huge amounts of CO_2 due to a reduction in the number of predators—is entirely fictitious, but if it hadn't been, the paper's argument would be scientifically valid.

Fake News

With its topical subject-matter, the previous spoof could almost be classed as "fake news". That currently fashionable term encompasses a number of well-established phenomena, from satire and political propaganda to the biased or partial reporting of real facts. All of these things have their counterparts in the world of physics.

Let's start with satire. That's all very well if the audience recognizes it as such, because then they will look below the surface to see the underlying message—and may even learn something from it. All too often, however, satire is taken at face value—resulting in the exact opposite of the desired effect. There were several examples in previous chapters where scientific spoofs intended to be humorous were mistaken for the real thing.

This is a particular problem on the internet, where there may be no obvious contextual hints that a piece is actually intended as satire. This principle is important enough to have a name—Poe's law—as Tom Chivers explained in the *Daily Telegraph* in 2009:

> Poe's Law states: "Without a winking smiley or other blatant display of humour, it is impossible to create a parody ... that someone won't mistake for the real thing." It was originally formulated by Nathan Poe in 2005 during a debate on christianforums.com about evolution, and referred to creationism [21].

A couple of creationism-related spoofs were mentioned in the "April Fool" chapter: the "Okay, We Give Up" editorial in *Scientific American*, and the bogus press release about a Bible-based value for pi. In accordance with Poe's Law, both of these have, on occasion, been taken as factual pieces.

One of the most remarkable examples of a scientific spoof being mistaken for the real thing occurred in 2001. Quoting from the *Daily Telegraph* again—this time from a news report in November that year:

> Documents found last week in an al-Qaeda safe house in Afghanistan that purport to be instructions on how to build a nuclear weapon were shown yesterday to be based on a spoof scientific article. The plans were discovered in a ruined house in Kabul after the Taliban fled the city, and included notes ostensibly showing how to create a nuclear device.

Bear in mind that this was just two months after the 9/11 atrocity—widely attributed to Al-Qaeda—so the idea that the organization possessed the plans for a nuclear weapon is a worrying one. Fortunately, however, the origin of those plans was a jokey periodical we met in the chapter on "Spoofs in Scientific Journals". As the *Daily Telegraph* goes on:

> The original, entitled "How To Build An Atomic Bomb In 10 Easy Steps" ... was one of a series run by the *Journal of Irreproducible Results*. Yesterday, a former editor, Marc Abrahams, confirmed that the documents, shown in a report by BBC reporter John Simpson, were from the article, though in a different format. "I have a copy of the issue and I have also seen footage of the BBC report and it is clear it is the same text," said Mr Abrahams [22].

The irony is that this appears to have happened by accident, as it's the sort of situation a government intelligence agency might go to great pains to create. The technical term for this is "disinformation", and it's a common practice in wartime (or during the build-up to war). One of the first exponents of the scientific variety of disinformation, during World War Two, was Professor R. V. Jones—the first physicist to be recruited by British military intelligence.[1] Looking back on that work later, Jones described yet another kind of "fake physics":

> Induced incongruities have a high place in warfare, where if the enemy can be induced to take incorrect action the war may be advantageously affected. A stratagem in which some of my wartime colleagues were involved is now well known as "the man who never was". These same colleagues also worked with me in some technical deceptions, of which one was the persuasion of the Germans in 1943 that our successes against the U-boats were due not to centimetric radar but to a fictitious infrared detector. We gained some valuable months while the Germans invented a beautiful anti-infrared paint and failed to find the true causes of their losses [23].

Disinformation exercises like Jones's infrared detector are the scientific equivalent of a second type of "fake news", propaganda. The scientific variety isn't particularly common outside wartime, although it does make an appearance in a few conspiracy theories. One of these, climate change denial, has already been mentioned. Another is the "9/11 truth" movement—the idea that, far from being the work of al-Qaeda, the destruction of the World Trade Centre was a carefully choreographed plot by the US government.

This is very much a fringe view, which is rarely given space in mainstream publications. If it was supported by research that appeared in a prestigious scientific journal, that would be big news indeed. Yet in 2016, claims surfaced on the internet that this had actually happened. Specifically, in the words of the Snopes fact-checking site:

> The *European Scientific Journal,* a peer-reviewed academic publication, concluded that the collapse of the Twin Towers and World Trade Centre Building 7 on 11 September 2001 was the result of a controlled demolition [24].

However, things were not quite as they appeared—as became clear when Snopes looked more carefully into the case. In effect, it was a case of "mistaken identity":

---

[1] "Military intelligence" is often taken to be an oxymoron; Jones was a notable exception to this.

In their July-August 2016 issue, the science news magazine *Europhysics News* (EPN) published a feature by a group of scientists who have long been involved with the promotion of 9/11 conspiracy theories. The piece argued that the structural failure of the WTC buildings on 9/11 was not adequately explained by burning jet fuel, and that it was instead better explained by a controlled demolition.

Despite the similarity of names, EPN is completely unrelated to the *European Scientific Journal*—and it's a news magazine, not an academic journal. That's not to imply that EPN is in any way disreputable; the magazine even put a disclaimer at the beginning of the 9/11 story to the effect that "this feature is somewhat different from our usual purely scientific articles, in that it contains some speculation". Nevertheless, the distinction was too subtle for many conspiracy theorists. As the Snopes article continues:

A variety of websites … published the claim that this news feature was a scientific article published in the *European Scientific Journal*, suggesting that because it was peer-reviewed it was a stronger validation than previously published conspiracy theories. In response, the publishers of the *European Scientific Journal* issued a statement clarifying that they had nothing to do with the article: "Regarding the recent developments on social media, we would like to inform the public that neither the *European Scientific Journal* nor the European Scientific Institute have published content on 9/11 attacks" [24].

Even in the mainstream media, there's a danger that scientific research may be misreported. This is rarely because of political bias (the commonest cause of "fake news" in other fields), but more often the result of pressure to make a story more "interesting" or "relevant" to a wide audience.

As an illustration of the constraints a scientific journalist has to work under, Martin Robbins wrote a spoof "Article about a Scientific Paper" for the *Guardian* website in September 2010. It begins as follows:

In this paragraph I will state the main claim that the research makes, making appropriate use of "scare quotes" to ensure that it's clear that I have no opinion about this research whatsoever.

In this paragraph I will briefly (because no paragraph should be more than one line) state which existing scientific ideas this new research "challenges".

If the research is about a potential cure, or a solution to a problem, this paragraph will describe how it will raise hopes for a group of sufferers or victims.

This paragraph elaborates on the claim, adding weasel-words like "the scientists say" to shift responsibility for establishing the likely truth or accuracy of the research findings onto absolutely anybody else but me, the journalist.

Robbins goes on in this vein for several more paragraphs, before apparently becoming alarmed by the thought that some readers might be losing interest. This prompts the following:

In this paragraph I will reference or quote some minor celebrity, historical figure, eccentric, or a group of sufferers; because my editors are ideologically committed to the idea that all news stories need a "human interest", and I'm not convinced that the scientists are interesting enough.

At this point I will include a picture, because our search engine optimisation experts have determined that humans are incapable of reading more than 400 words without one.

The picture in question, showing a triceratops dinosaur superimposed on a spiral galaxy, is captioned: "this picture has been optimised by SEO experts to appeal to our key target demographics". A few more paragraphs follow, then Robbins winds up with "the final paragraph will state that some part of the result is still ambiguous, and that research will continue" [25].

The problem with this "populist" approach is that it can emphasize outlandish speculations at the expense of more sober ones. A case in point was the astronomical object named 'Oumuamua, which was observed passing through the solar system in late 2017. It was the first object ever seen in the vicinity of Earth that was travelling on a trajectory that originated in interstellar space. To certain people—sci-fi fans and UFO buffs, for example—that could only mean one thing: 'Oumuamua was an alien spacecraft.

The facts, however, were not in favour of this. 'Oumuamua was much more likely to be an inert comet or asteroid, because its trajectory conformed almost perfectly to what would be expected from an object coasting freely under the action of the Sun's gravity. There was a tiny non-gravitational acceleration, but this could adequately be accounted for by comet-like outgassing.

The problem arose a year later, when Shmuel Bialy and Avi Loeb of the prestigious Harvard-Smithsonian Centre for Astrophysics wrote a paper entitled "Could Solar Radiation Pressure Explain 'Oumuamua's Peculiar Acceleration?"

The authors' answer to their own question was "yes, it might do"—depending on the exact shape of the object, which remains unknown. However, they went a step further than that. Towards the end of the paper, they wrote:

> Considering an artificial origin, one possibility is that 'Oumuamua is a lightsail, floating in interstellar space as debris from an advanced technological equipment … A more exotic scenario is that 'Oumuamua may be a fully operational probe sent intentionally to Earth vicinity by an alien civilization [26].

A light-sail, or solar sail, is a proposed type of space vehicle that is propelled by radiation pressure from the Sun or other stars (see Fig. 4).

There's nothing wrong, per se, with injecting a little speculation into a scientific paper. If you conclude that 'Oumuamua's trajectory might have been affected by radiation pressure, then there shouldn't be any harm in mentioning the "light sail" idea—because your primary audience, in the form of fellow academics, will view this in the context of a range of other, more likely, hypotheses.

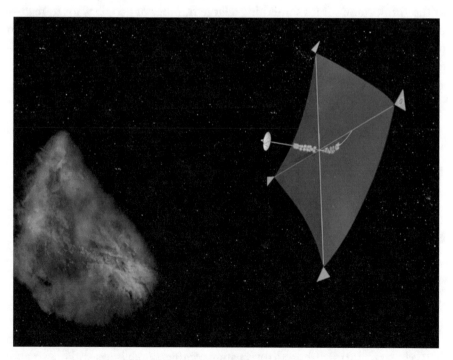

Fig. 4 Artist's impression of a space probe propelled by a solar sail—something 'Oumuamua almost certainly wasn't, despite media reports (NASA image)

Unfortunately, once an idea is in the public domain, it won't just be seen by academics. If it's an eyecatching one, it will be picked up by journalists—and they have a natural homing instinct for "what people want to hear". The public isn't interested in cometary outgassing or radiation pressure—but it is fascinated by alien spaceships. Putting an alien spaceship in a scientific paper is an open invitation to the media to tell the world about it. To quote the Snopes website:

> If you are a journalist looking for a sensational claim to attribute to a scientist, your best bet would be to set your sights on the penultimate paragraph of a scientific paper … Unfortunately, presenting this portion of a study as a headline—while great for clicks—rarely elucidates the actual scientific debate at hand and often muddies the factual information presented by the paper. A prime example of this phenomenon hit the viral news machine in early November 2018 with headlines such as "Mysterious Interstellar Object Floating in Space Might Be Alien, Say Harvard Researchers" running in *USA Today* [27].

This comes back to the problem mentioned earlier—that science reporting is habitually twisted by the need to focus on the "interesting" at the expense of the "important". If it's something a large number of people want to read—like the possibility of an alien visitation—it will be repeated *ad nauseam* in preference to other, more likely explanations. As a *Gizmodo* article (appropriately titled "No, 'Oumuamua Is Probably Not an Alien Spaceship") put it:

> Other scientists were sceptical of the paper. "It's important not to take advantage of your institution's brand to over-amplify results that are unverified or highly speculative," Chanda Prescod-Weinstein, assistant professor of physics at the University of New Hampshire, told *Gizmodo*. "It doesn't just affect the department's reputation but also the rest of the field." [28]

References

1. J. Stribling, M. Krohn, D. Aguayo, SCIgen: an automatic CS paper generator. https://pdos.csail.mit.edu/archive/scigen/
2. Prank Fools US Science Conference. BBC News (April 2005). http://news.bbc.co.uk/1/hi/world/americas/4449651.stm
3. R. Van Noorden, Publishers withdraw more than 120 Gibberish papers. Nature (February 2014). https://www.nature.com/news/publishers-withdraw-more-than-120-gibberish-papers-1.14763

4. E. Hunt, Nonsense paper written by iOS autocomplete accepted for conference. The Guardian (October 2016). https://www.theguardian.com/science/2016/oct/22/nonsense-paper-written-by-ios-autocomplete-accepted-for-conference

5. P. Aldhous, CRAP paper accepted by journal. New Scientist (June 2009). https://www.newscientist.com/article/dn17288-crap-paper-accepted-by-journal/

6. D. Butler, Investigating journals: the dark side of publishing. Nature (March 2013). https://www.nature.com/news/investigating-journals-the-dark-side-of-publishing-1.12666

7. M. Laakso, B.-C. Björk, Anatomy of open access publishing. BMC Medicine **10**, 124–132 (2012)

8. Neuroskeptic, Predatory journals hit by star wars sting. Discovermagazine.com (July 2017). http://blogs.discovermagazine.com/neuroskeptic/2017/07/22/predatory-journals-star-wars-sting/

9. Z. Faulkes, Stinging the predators. Version 8.2 (October 2018). https://figshare.com/articles/Stinging_the_Predators_A_collection_of_papers_that_should_never_have_been_published/5248264

10. N.T. Redd, Fake science paper about star trek and warp 10 was accepted by predatory journals. Space.com (February 2018). https://www.space.com/39672-fake-star-trek-science-paper-published.html

11. Memory alpha episode summary for "Threshold". http://memory-alpha.wikia.com/wiki/Threshold_(episode)

12. M. Gault, Scientist published papers based on *Rick and Morty* to expose predatory academic journals. vice.com (October 2018). https://motherboard.vice.com/en_us/article/3km5j8/scientist-published-papers-based-on-rick-and-morty-to-expose-predatory-academic-journals

13. M. Lynch, Is a science peace process necessary? *The One Culture* (University of Chicago Press, 2001), pp. 48–60

14. Z. Sardar, *Thomas Kuhn and the Science Wars* (Icon Books, Cambridge, 2000), pp. 4–7

15. A.D. Sokal, Transgressing the boundaries: towards a transformative hermeneutics of quantum gravity. Social Text #46/47 (Spring/Summer 1996). https://physics.nyu.edu/faculty/sokal/transgress_v2/transgress_v2_singlefile.html

16. R. Sheldrake, In the Vale of soul-making. https://www.sheldrake.org/about-rupert-sheldrake/interviews/in-the-vale-of-soul-making

17. A.D. Sokal, A physicist experiments with cultural studies. Lingua Franca (May/June 1996). https://physics.nyu.edu/faculty/sokal/lingua_franca_v4/lingua_franca_v4.html

18. J. Delingpole, Do penises cause climate change? The Spectator (June 2017). https://www.spectator.co.uk/2017/06/do-penises-cause-climate-change-discuss/

19. Hoax bacteria study tricks climate sceptics. Reuters (8 November 2007). https://uk.reuters.com/article/environment-climate-hoax-dc/hoax-bacteria-study-tricks-climate-skeptics-idUKL0887458220071108

20. D.A. Klein, et al., Carbon dioxide production by benthic bacteria. researchgate. net. https://www.researchgate.net/publication/241480752_Carbon_dioxide_ production_by_benthic_bacteria_the_death_of_manmade_global_ warming_theory

21. T. Chivers, Internet rules and laws. The Daily Telegraph (23 October 2009). https://www.telegraph.co.uk/technology/news/6408927/Internet-rules-and-laws-the-top-10-from-Godwin-to-Poe.html

22. R. Highfield, Al-Qaeda's atom plans were spoof science. Daily Telegraph (November 2001). https://www.telegraph.co.uk/news/worldnews/asia/afghanistan/1362926/Al-Qaedas-atom-plans-were-spoof-science.html

23. R.V. Jones, The theory of practical joking: its relevance to physics, in *A Random Walk in Science*, ed. by R. L. Weber, E. Mendoza, (Institute of Physics, Bristol, 1973), pp. 8–14

24. A. Kasprak, Did a European Scientific Journal conclude 9/11 was a controlled demolition? Snopes.com (October 2016). https://www.snopes.com/fact-check/journal-endorses-911-conspiracy-theory/

25. M. Robbins, This is a news website article about a scientific paper. The Guardian (September 2010). https://www.theguardian.com/science/the-lay-scientist/2010/sep/24/1

26. S. Bialy, A. Loeb, Could solar radiation pressure explain 'Oumuamua's peculiar acceleration? Astrophys. J. Lett. (November 2018). https://arxiv.org/abs/1810.11490

27. A. Kasprak, The science behind those viral Oumuamua alien spacecraft stories. Snopes.com (November 2018). https://www.snopes.com/news/2018/11/07/the-science-behind-those-viral-oumuamua-alien-spacecraft-stories/

28. R.F. Mandelbaum: No, Oumuamua is probably not an alien spaceship. Gizmodo (November 2018). https://gizmodo.com/no-oumuamua-is-probably-not-an-alien-spaceship-1830255239

Thinking Outside the Box

Abstract This final chapter stretches the definition of "fake physics" to look at a few examples that convey a serious scientific message. We start with some sci-fi-sounding topics that are nonetheless amenable to proper scientific analysis, and then go on to some well-known "thought experiments"—again addressing serious scientific subjects by means of science-fictional scenarios. These range from black holes and relativistic time dilation—which are impractical to study in the real world—to the question of other universes governed by completely different physical laws. The chapter concludes with an overview of the whole spectrum of fake physics encountered in the course of the book.

Science Fact Posing as Science Fiction?

The first chapter of this book was called "Science Fiction Posing as Science Fact", and it presented several amusing examples—such as Isaac Asimov's "Thiotimoline" spoofs—of exactly that. But does the opposite ever happen—does science fact ever pose as science fiction?

Taking the question in a literal sense, the answer is obviously "no". On the other hand, there are numerous cases where serious scientists have applied valid scientific methods to subjects that are more commonly associated with SF. There were several examples of this in the chapter on "The Relativity of Wrong", ranging from wormholes and tachyons to the "many worlds" hypothesis and Fermi's paradox.

The last of these, formulated in the middle of the twentieth century by the Nobel-prize winning physicist Enrico Fermi, deals with the archetypal SF topic of extraterrestrial civilizations. Surprisingly, exactly the same subject was touched on much earlier, in the last year of the nineteenth century, by a Nobel Prize winner of the previous generation, Max Planck.

© Springer Nature Switzerland AG 2019
A. May, *Fake Physics: Spoofs, Hoaxes and Fictitious Science*, Science and Fiction,
https://doi.org/10.1007/978-3-030-13314-6_7

Planck is best known as the "father" of quantum theory, who introduced the fundamental quantum of action now referred to as Planck's constant. This is one of a small number of universal constants—others being the speed of light, the gravitational constant, the Coulomb constant of electromagnetism and the Boltzmann constant of thermodynamics.

What Planck did in 1899 was to propose a "universal" system of measurement—now known as Planck units—in which all these constants are exactly equal to 1. This was a purely hypothetical exercise, since the resulting units turn out to be totally impractical (for example the Planck length is of the order of 10^{-35} metres, and the Planck temperature circa 10^{32} degrees Celsius). Nevertheless, the resulting units are far less arbitrary than existing ones, as Planck explained in the following way:

> All the physical measurement systems hitherto used, including the so-called absolute centimetre-gram-second system, owe their origin to the coincidence of random circumstances, in that the choice of units on which each system is based does not depend on general points of view that are necessary for all places and times, but rather on the special needs of our terrestrial culture. Thus, the units of length and time have been derived from the present dimensions and motion of our planet, and the units of mass and temperature from the density and fundamental points of water, as a liquid at the Earth's surface.

Planck's system, on the other hand, has no ties to the Earth at all—and this where he made explicit reference to "aliens":

> There is the possibility of establishing units of length, mass, time and temperature which, independent of specific bodies or substances, retain their meaning for all cultures, even extraterrestrial and non-human ones [1].

Since Planck's time, there has been increasing divergence between the way alien life is visualized by scientists, on the one hand, and by the sci-fi and UFO community on the other. In the former case, the aliens are genuinely "alien", of a kind that might reasonably evolve on a different planet, while the latter persists in viewing aliens as similar to Earthly humans in appearance and behaviour.

Most of the "evidence" put forward to support the latter view is anecdotal, and thus not amenable to scientific analysis. A notable exception was the so-called "Face on Mars"—a topographic feature photographed by the first Mars orbiters which, in early low resolution images, bore a marked resemblance to a human face.

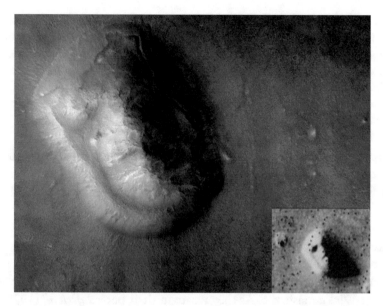

Fig. 1 High-resolution image of the so-called "Face on Mars" by the Mars Reconnaissance Orbiter, with the original, low resolution image—as studied by Mark Carlotto—shown in the inset (NASA images)

Its advocates claimed that was exactly wat it was—a huge sculpture carved by an ancient humanoid civilization on Mars. Most scientists, however, considered it to be a natural formation with a chance resemblance to a face—something that higher resolution imagery has made much more obvious (see Fig. 1).

Excitedly claiming that the face-like formation was "obvious" evidence for an ancient Martian civilization, as many enthusiasts did, was unscientific. Unlike most such claims, however, this one was open to scientific examination—because the NASA imagery of the face-like object existed as hard data.

What Mark Carlotto of the Analytic Sciences Corporation did, in a study written up in the peer-reviewed journal *Applied Optics* in 1988, was to apply standard image-processing techniques to those original, low resolution NASA photographs. Here is the abstract of his paper:

Image processing results in support of ongoing research into the origin of a collection of unusual surface features on Mars are presented. The focus of the investigation is on a mile long feature in the Cydonia region of Mars which resembles a humanoid face that was imaged by Viking orbiter in 1976… Image enhancements of the face show it to be a bisymmetrical object having two eyes, a nose, and a mouth; fine structure in the mouth suggesting teeth are apparent in the enhanced imagery as well as crossed symmetrical lines on the forehead. Facial features are also evident in the underlying 3D surface which was reconstructed using a single image shape-from-shading technique. Synthetic images derived from the 3D model by computer graphics techniques suggest that the impression of facial features evident in the original Viking imagery are not a transient phenomenon, i.e. they persist over a wide range of illumination and viewing conditions [2].

Surprisingly, Carlotto's work provided scientific support for the idea that the formation "looked like a face". It was only when later spacecraft, such as Mars Reconnaissance Orbiter, acquired better pictures of the area that the illusion disappeared.

Another "extraterrestrial" topic that crosses the boundary between pseudo-science and real science is panspermia: the hypothesis that the seeds of life came to Earth from outer space. The idea had been around for a long time, but it was popularized in the 1970s by maverick scientists Fred Hoyle and Chandra Wickramasinghe. As they wrote in their book *Lifecloud* (1978):

Our argument is that life arrived eventually on Earth by being showered as already living cells from comet-type bodies [3].

Although the pair were outwardly mainstream scientists—Hoyle spent most of his career at Cambridge University, while Wickramasinghe was a professor at the University of Wales—their claims about panspermia were largely ignored by the rest of the science community. Nevertheless, the theory doesn't fall in the "not even wrong" category: it makes a number of falsifiable claims, such as the ability of certain types of living organism to survive in the vacuum of space.

This issue was addressed in a paper called "Can spores survive in interstellar space?" by Peter Weber and J. Mayo Greenberg of the University of Leiden, which was published in that most prestigious of all scientific journals, *Nature*, in 1985. The paper's abstract reads as follows:

Inactivation of spores (*Bacillus subtilis*) has been investigated for the first time in the laboratory by vacuum ultraviolet radiation in simulated interstellar conditions. Remarkably, damage produced at the normal interstellar particle temperature of 10 K is less than at higher temperatures, the major damage being produced by radiation in the 2000–3000 angstrom range. Our results place constraints on the panspermia hypothesis [4].

This could be considered a partial debunking of Hoyle and Wickramasinghe's hypothesis—but it certainly doesn't kill it off as definitively as its critics might have liked. For a real hatchet job, it's necessary to turn to an even more "fringe" topic—extrasensory perception (ESP)—and another paper that appeared in *Nature* a few years earlier. This was "Is There Any Scientific Explanation of the Paranormal?" (1979) by J. G. Taylor and E. Balanovski of King's College in London. Here is the abstract:

The apparent impossibility of the occurrence of "paranormal" phenomena has not discouraged their extensive investigation, although there has not been any uniformly accepted validation or explanation by the scientific community. To clarify exactly how difficult ESP phenomena are to explain, it is necessary to place them in the framework of modern science. Explanations of the phenomena have been brought forward which have been claimed to make them more respectable. These explanations must also be looked at from the point of view of modern science and this paper is devoted to that task. In particular we wish to indicate that on theoretical grounds the only scientifically feasible explanation could be electromagnetism (EM) involving suitably strong EM fields. Thus we regard that this paper completes our earlier work where we presented experimental results giving the level of the EM signals emitted by subjects when engaged in supposedly paranormal activity. These EM levels were many orders of magnitude lower than the ones we calculate here as needed to achieve paranormal effects. Taken together the two papers are a strong argument against the validity of the paranormal [5].

Another archetypal SF theme that occasionally crops up in respectable science journals is time travel. One of the first papers to address the subject was the cryptically titled "Rotating Cylinders and the Possibility of Global Causality Violation", by Frank J. Tipler of the University of Maryland, which appeared in *Physical Review* in 1974. Its abstract is less coy than the title with regard to the subject-matter:

In 1936 van Stockum solved the Einstein equations $G_{\mu\nu} = -8\pi T_{\mu\nu}$ for the gravitational field of a rapidly rotating infinite cylinder. It is shown that such a field violates causality, in the sense that it allows a closed time-like line to connect any two events in spacetime. This suggests that a finite rotating cylinder would also act as a time machine [6].

Surprisingly, Tipler's paper isn't the only work to bear the title "Rotating Cylinders and the Possibility of Global Causality Violation". That's also the title of a short story, written a few years later, by the SF author Larry Niven. Here is what his official online bibliography says about the story:

First published in *Analog*, August 1977. Niven borrowed the title of a mathematics paper by Frank J. Tipler for this look at the principle of "cosmic censorship", the way the universe protects itself (sometimes rather violently) from the paradoxes implied by time travel [7].

Science fiction aside, the important question is whether Tipler's time machine would work in the real world. The answer is that it might well do—if only it could be built. Unfortunately, from a practical point of view that's virtually impossible. A "Tipler cylinder" would need to have the density of a neutron star, more than a trillion times that of normal matter. And because neutron stars are spherical, you would need to line up several of them to make a cylinder. To quote Brian Clegg, from his book *How to Build a Time Machine*:

> So the challenges facing interstellar engineers wanting to make a Tipler cylinder are, to say the least, nontrivial. They have to locate at least ten neutron stars and drag them together... This requires travel over vast distances, plus the ability to manipulate something the weight of a star from place to place across tens or hundreds of light years. We would then need to force ten of them together, equalize their rotation, and spin them up to maybe three times the revs. Finally, we would have to apply some massive force, probably an antigravitational force, to keep the stars in a cylinder—and we would have to have some way (again we're talking antigravity) to protect our time travellers from being dragged apart by tidal forces around the cylinder. All in all, Tipler's cylinders are a nice idea ... but it isn't going to happen [8].

The fact is that Tipler didn't postulate his time-bending cylinders as a viable engineering proposition, but as a "thought experiment"—something that has become increasingly common in physics as the subject has strayed further and further from the everyday world. In their own way, thought experiments are yet another kind of "fake physics"—and they're what we're going to look at now.

Thought Experiments

Here is how physicist Jim al-Khalili introduces the subject of thought experiments in his 2013 book *Paradox*:

> When faced with difficulties in testing the predictions of their theories, physicists sometimes resort to what are called "thought experiments"—idealized imaginary scenarios that do not violate any laws of physics and yet are too impractical or hypothetical to set up as real experiments in the lab [9].

One of the first—and still one of the most famous—of all thought experiments concerns a microscopic creature called "Maxwell's demon". It was devised in the middle of the nineteenth century by the Scottish physicist James Clerk Maxwell—a pioneer in, among other things, the "moving molecule" theory of thermodynamics. The second law of thermodynamics famously states that it's impossible to transfer heat from a colder body to a hotter one without the expenditure of energy. Maxwell's thought experiment shows how this might be achieved with the aid of his hypothetical demon (see Fig. 2).

Here is Isaac Asimov's summary of Maxwell's thought experiment (note that his wording flips left and right compared to the example in Fig. 2):

> If two containers of gas at the same temperature were connected by a tiny door guarded by a tiny demon, one could imagine that door being opened whenever a slowly moving molecule was passing to the right, but not the left; or whenever a quickly moving molecule was passing to the left, but not the to the right. In this way the fast molecules would accumulate in the left flask, which would thus grow hotter and hotter, while the slow molecules would accumulate in the right flask, which would grow colder and colder. Heat would flow in this fashion continuously from cold to hot in defiance of the second law [10].

Although the scenario may sound flippant, it's actually very serious. Out of all the physical "laws" devised by scientists over the years, the second law of thermodynamics is the one that most of them would consider least susceptible to amendment. It's virtually an article of faith that any theory or proposition that violates the second law is going to be proved wrong. As the twentieth century physicist and science popularizer Arthur Eddington put it:

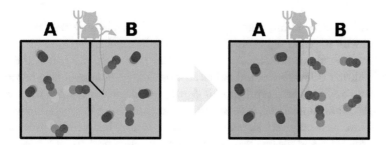

Fig. 2 The "Maxwell's demon" thought experiment. By controlling the opening and closing of a door between two containers, the demon makes the left one colder and the right one hotter (Wikimedia user Htkym, CC-BY-SA-3.0)

If your theory is found to be against the second law of thermodynamics I can give you no hope; there is nothing for it but to collapse in deepest humiliation [11].

This means there's almost certainly some flaw in the "Maxwell's demon" scenario—but it's far from being an obvious one. In fact it hasn't been pinned down to everyone's satisfaction even today. Quoting Jim al-Khalili again:

Physicists today have chased the demon all the way down to the quantum realm and the strange rules that operate at atomic scales… Heisenberg's uncertainty principle states that we can never know exactly both where a particle (or air molecule) is and at the same time exactly how fast it is moving; there is always a kind of fuzziness. And it is this fuzziness, many argue, that is ultimately needed to preserve the Second Law of Thermodynamics [12].

Another branch of physics where thought experiments abound is Einstein's theory of relativity—both "special" (dealing with very high speeds, close to that of light) and "general" (dealing with very strong gravitational fields, such as those created by black holes). By the nature of the regimes in question, relativistic thought-experiments are often distinctly science-fictional in tone—a point made by Stephen Hawking in one of his last lectures:

It is said that fact is sometimes stranger than fiction, and nowhere is that more true than in the case of black holes. Black holes are stranger than anything dreamed up by science fiction writers, but they are firmly matters of science fact [13].

In his most famous book, *A Brief History of Time*, Hawking's dramatization of how a black hole forms sounds more like something out of a sci-fi novel than a physics textbook:

Suppose an intrepid astronaut on the surface of a collapsing star, collapsing inward with it, sent a signal every second, according to his watch, to his spaceship orbiting about the star. At some time on his watch, say 11:00, the star would shrink below the critical radius at which the gravitational field becomes so strong nothing can escape, and his signals would no longer reach the spaceship. As 11:00 approached, his companions watching from the spaceship would … have to wait only very slightly more than a second between the astronaut's 10:59.58 signal and the one that he sent when his watch read 10:59.59, but they would have to wait forever for the 11:00 signal… The time interval between the arrival of successive waves at the spaceship would get longer and longer, so the light from the star would appear redder and redder and fainter and fainter. Eventually, the star would be so dim that it could no longer be seen from the spaceship; all that would be left would be a black hole in space [14].

Another famous thought experiment occurs in special relativity, in the form of the so-called "twin paradox". Once again this involves a highly science-fictional scenario, as Jim al-Khalili explains:

The storyline of this paradox may sound like science fiction, but it is in fact perfectly within the mainstream science taught to every physics student as an example of the implications of relativity, even if it is not technologically achievable just yet. It involves a spacecraft capable of reaching near light speed—which, while we have no means of developing such a craft at the moment, is nevertheless perfectly admissible in principle [15].

At the heart of the twin paradox is the phenomenon of time dilation: the true but counter-intuitive fact that time runs more slowly for someone on a fast-moving spaceship than for everyone else back on Earth. The "paradox" lies in the fact that speed can only ever be measured in a relative fashion. This means that, when the spaceship is coasting along at its maximum speed, an astronaut on board could validly claim to be at rest and that it was the planet Earth that was in rapid motion.

The twin paradox imagines a set of identical twins[1]—one who stays on Earth and one who goes on a space journey at close to the speed of light. When the travelling twin returns to Earth, one of the two will have aged more than the other. But which one? If motion is purely relative, it could be either of them.

As it happens, Einstein's theory provides a definitive resolution to the paradox: it's the stay-at-home twin who ages more than the spacefaring one. The situation isn't as symmetric as it sounds, because the travelling twin has to undergo periods of acceleration and deceleration that the Earthbound twin doesn't experience.

Ironically, despite the fact that time dilation is so often explained in science-fictional terms, it rarely features in science fiction itself. People who learn all their astrophysics from *Star Wars* and *Star Trek* may never have heard of it. One movie that does hint at it, however, is *Planet of the Apes* (1968). Near the beginning, Charlton Heston's character Taylor dictates the following log entry:

---

[1] In the interests of balance, many modern textbooks take them to be non-identical male/female twins – but that weakens the thought experiment, which is more striking if the twins are literally identical to start with. So let's be old-fashioned and imagine they are both female.

In less than an hour we'll finish our six months out of Cape Kennedy. Six months in deep space—by our time, that is. According to Dr Hasslein's theory of time in a vehicle travelling nearly the speed of light, the Earth has aged nearly 700 years since we left it, while we've aged hardly at all... The men who sent us on this journey are long since dead and gone [16].

It's odd that the theory is ascribed to a fictional "Dr Hasslein" rather than the real-world Einstein. Perhaps that's because the source novel, by Pierre Boulle, doesn't mention Einstein by name either. Nevertheless, it's obviously Einstein's theory that is being paraphrased by one of Boulle's characters in the following quote:

While we are moving at this speed, our time diverges perceptibly from time on Earth, the divergence being greater the faster we move. At this very moment, since we started this conversation, we have lived several minutes which correspond to a passage of several months on our planet [17].

Even in written SF—which is traditionally more faithful to science than its Hollywood counterpart—time dilation only tends to feature when it is central to the plot. Examples are the novels *Return to Tomorrow* (1954) by L. Ron Hubbard (better known as the founder of scientology), *Time for the Stars* (1956) by Robert A. Heinlein and *The Forever War* (1974) by Joe Haldeman.

Hubbard's novel even goes so far as to quote the correct mathematical formula for time dilation:

$$T_v = T_0 \cdot \sqrt{1 - \frac{v^2}{c^2}}$$

where T_v is time as measured on a spaceship moving at speed v, T_0 is time as measured on Earth, and c is the speed of light [18].

The reason that time dilation seems so counter-intuitive is that even the highest speeds encountered in the real world are much less than c, which is around 300,000 kilometres per second. Under these circumstances, the difference between T_v and T_0 is negligibly small. It would be a different matter altogether if the value of c was much smaller—which brings us to the subject of the next section.

Different Physics

The cosmologist George Gamow was mentioned in the "Spoofs in Science Journals" chapter, as the instigator of the famous "alpha, beta, gamma" paper. Another example of Gamow's whimsical sense of humour is his fictional creation Mr Tompkins, who appeared in a series of books starting with *Mr Tompkins in Wonderland* in 1939.

These books take the idea of "thought experiments" a stage further, to the point where the constants of nature have different values from their real-world ones. Gamow had a didactic purpose in doing this, because it makes the exotic phenomena of relativity and quantum physics much more obvious in everyday situations. As he wrote in the introduction to *Mr Tompkins in Paperback*:

> The deviations between the common notions and those introduced by modern physics are, however, negligibly small so far as the experience of ordinary life is concerned. If, however, we imagine other worlds, with the same physical laws as those of our own world, but with different numerical values for the physical constants determining the limits of applicability of the old concepts, the new and correct concepts of space, time and motion, at which modern science arrives only after very long and elaborate investigations, would become a matter of common knowledge… The hero of the present stories is transferred, in his dreams, into several worlds of this type, where the phenomena, usually inaccessible to our ordinary senses, are so strongly exaggerated that they could easily be observed as the events of ordinary life [19].

Mr Tompkins's very first adventure concerns the theory discussed at the end of the last section: special relativity. As mentioned there, relativistic effects are negligible in the real world—but Mr Tompkins dreams that he is in a parallel world where the speed of light is much slower. His first surprise concerns a relativistic effect that hasn't been mentioned yet: "length contraction", whereby objects moving at a different speed to the observer appear foreshortened (see Fig. 3).

Here is an excerpt from the story itself:

> A single cyclist was coming slowly down the street and, as he approached, Mr Tompkins's eyes opened wide with astonishment. For the bicycle and the young man on it were unbelievably shortened in the direction of the motion, as if seen through a cylindrical lens. The clock on the tower struck five, and the cyclist, evidently in a hurry, stepped harder on the pedals. Mr Tompkins did not notice that he gained much in speed, but, as the result of his effort, he shortened still

Fig. 3 Mr Tomkins dreams he is in a world where light travels much more slowly, so that cyclists appear foreshortened—or, when he himself is the cyclist—everything else appears foreshortened (public domain images)

more and went down the street looking exactly like a picture cut out of cardboard.

Before long, Mr Tompkins acquires his own bicycle and races off on it:

He expected that he would be immediately shortened, and was very happy about it as his increasing figure had lately caused him some anxiety. To his great surprise, however, nothing happened to him or to his cycle. On the other hand, the picture around him completely changed. The streets grew shorter, the windows of the shops began to look like narrow slits, and the policeman on the corner became the thinnest man he had ever seen. "'By Jove!" exclaimed Mr Tompkins excitedly, "I see the trick now. This is where the word relativity comes in. Everything that moves relative to me looks shorter for me, whoever works the pedals!"

When he arrives at the railway station, he witnesses the effects of time dilation at first hand:

A gentleman obviously in his forties got out of the train and began to move towards the exit. He was met by a very old lady, who, to Mr Tompkins's great surprise, addressed him as "dear Grandfather". This was too much for Mr Tompkins. Under the excuse of helping with the luggage, he started a conversation. "Excuse me, if I am intruding into your family affairs," said he, "but are you really the grandfather of this nice old lady? You see, I am a stranger here, and I never –"

"Oh, I see," said the gentleman, smiling with his moustache… "The thing is really quite simple. My business requires me to travel quite a lot, and, as I spend most of my life in the train, I naturally grow old much more slowly than my relatives living in the city. I am so glad that I came back in time to see my dear little grand-daughter still alive!" [20]

As mentioned at the start of this chapter, in the context of Max Planck and his "alien-friendly" measurement system, the speed of light is one of several fundamental constants that define the way universe works. Others relate to the strength of gravity, electromagnetism and the "strong" and "weak" nuclear forces. Surprisingly, if any of these constants were slightly different, the universe as we know it couldn't exist. As physicist Michio Kaku explains:

If the strong nuclear force were a bit weaker, nuclei like deuterium would fly apart, and none of the elements of the universe could have been successively built up in the interiors of stars via nucleosynthesis. If the nuclear force were a bit stronger, stars would burn their nuclear fuel too quickly… If the weak force were a bit weaker, neutrinos would interact hardly at all, meaning that supernovae could not create elements beyond iron. If the weak force were a bit stronger, neutrinos might not escape properly from a star's core, again preventing the creation of the higher elements.

The fact that the universe appears to be "fine-tuned" in this way—with no obvious reason why the various constants have the values they do—is seen as a puzzle by many people. It's a puzzle that takes on philosophical dimensions when you realize that without this fine tuning, human life could not exist.

Some people take this as evidence for a benign and intelligent creator of the universe. That's a rather short-sighted view, however. You might as well say that the Earth's favourable position in the Solar System—not too close to the Sun and not too far away—is likewise the work of an intelligent creator.

Of course, we know there are other planets in the Solar System—and many others around other stars—so that in a statistical sense, the favourable conditions on Earth are bound to arise somewhere. By analogy, our universe may just by one of a large number—the one in which "life as we know it" happens to be possible. To quote Michio Kaku again:

Other scientists, like Sir Martin Rees of Cambridge University, think that these cosmic accidents give evidence for the existence of the multiverse. Rees believes that the only way to resolve the fact that we live within an incredibly tiny band of hundreds of coincidences is to postulate the existence of millions of parallel universes. In this multiverse of universes, most universes are dead. The proton is not stable. Atoms never condense. DNA never forms. The universe collapses prematurely or freezes almost immediately [21].

There's a difference here from the "multiple universes" hypothesized in the many-worlds interpretation of quantum physics, and mentioned in the chapter on "The Relativity of Wrong". Those universes differ in detail—diverging from each other every time a quantum wave function collapses—but the laws of physics are the same in all of them.

By contrast, the multitude of universes postulated by scientists like Martin Rees all have different laws of physics. That's not as familiar an idea as the "branching realities" of the many worlds hypothesis—which have given rise to countless sci-fi stories—but it's still a profitable one for the more adventurous writers of SF.

One of the first authors to grapple seriously with the subject was Isaac Asimov, in his 1973 novel *The Gods Themselves*. This recounts the events that follow the discovery of an "impossible" substance, plutonium-186, with 94 protons and just 92 neutrons in a nucleus.

To understand why that is so impossible—in our universe—it's necessary to take a closer look at the nucleus of an atom. This is made up of protons and neutrons, the former with a positive electrical charge and the latter electrically neutral. Left to themselves, a cluster of protons would be flung apart by electrostatic repulsion. This repulsion can only be counterbalanced by an attraction—and that's provided by the strong nuclear force, which acts between both protons and neutrons.

It turns out that, in order for the strong force to overcome electrical repulsion, an element like plutonium needs significantly more neutrons than protons in its nucleus. In our universe, the closest thing to a stable isotope of plutonium is plutonium-244, with no fewer than 150 neutrons in its nucleus—more than half again as many as the mysterious substance in Asimov's novel. Wherever that came from, it wasn't our universe. As one of the characters says:

> We are faced with a substance, plutonium-186, that cannot exist at all, let alone as an even momentarily stable substance, if the natural laws of the universe have any validity at all. It follows, then, that since it does indubitably exist and did exist as a stable substance to begin with, it must have existed, at least to begin with, in a place or at a time or under circumstances where the natural laws of the universe were other than they are. To put it bluntly, the substance we are studying did not originate in our universe at all, but in another—an alternate universe—a parallel universe. Call it what you want.

A bit later, the same character expands on this point:

> We cannot say in how many different ways the laws of the para-universe differ from our own, but we can guess with some assurance that the strong nuclear interaction, which is the strongest known force in our universe, is even stronger in the para-universe; perhaps a hundred times stronger. This means that protons are more easily held together against their own electrostatic attraction and that a nucleus requires fewer neutrons to produce stability [22].

In Asimov's novel, there's no way of actually travelling to the "para-universe". Even if it were possible, humans could never survive in a place where the strong nuclear force was so different, because it would change the whole chemistry on which we're based.

The situation would be less catastrophic if it was the force of gravity, rather than the strong nuclear force, that was altered. Of course we feel the effect of gravity, but it doesn't play a fundamental role in our constitution in the way that nuclear forces and electromagnetism do.

In his far-fetched—but still firmly science-based—novel *Raft* (1991), Stephen Baxter imagined what life would be like for humans if they were suddenly transported to a universe with a much stronger force of gravity. It's based on an earlier short story of the same title, about which Baxter wrote:

> The short story "Raft" came from a throwaway piece of speculation I read on the fine-tuning of physical parameters in our universe. If gravity were a little stronger, stars would be smaller and would burn out more quickly [23].

In the story itself, a character explains that "five hundred years ago a great warship—chasing some forgotten opponent—blundered through a portal … it left its own universe and arrived here". But where is "here", exactly? As the same character explains later:

> The force of gravity is a billion times stronger here than in the universe we came from… Even an object as puny as a man has a significant gravity field… If the Solar System were moved here, the Sun's increased pull would whip Earth round its orbit in 17 minutes.

Some of the effects of stronger gravity, like the one just mentioned, are obvious. Others are more subtle—not to mention alarming, such as the fact that a newly formed star only shines for a few years:

> The stars in the nebula shine by burning hydrogen. When they die, after a year or so, they leave more complex substances behind. The stars keep us alive. They give us light and warmth … but our nebula is running out of hydrogen. Another few years and no more stars [23].

The subject of different universes, with different physical laws, has been addressed by real scientists as well as SF authors. For example, Fred Adams of the University of Michigan wrote a paper called "Stars in Other Universes: Stellar Structure with Different Fundamental Constants", which was published in the *Journal of Cosmology and Astroparticle Physics* in 2008.

Adams's conclusions are really quite upbeat—not exactly demolishing the "fine-tuning" argument, but at least weakening it. Quoting from his paper:

In this paper, we have developed a simple stellar structure model to explore the possibility that stars can exist in universes with different values for the fundamental parameters that determine stellar properties. This paper focuses on the parameter space given by ... the gravitational constant, the fine structure constant, and a composite parameter that determines nuclear fusion rates. The main result of this work is a determination of the region of this parameter space for which bona fide stars can exist. Roughly one fourth of this parameter space allows for the existence of "ordinary" stars. In this sense, we conclude that universes with stars are not especially rare (contrary to previous claims).

He goes on to consider what happens in the "three quarters" of universes where normal stars are impossible:

For universes where no nuclear reactions are possible, we have shown that unconventional stellar objects can fill the role played by stars in our universe, i.e. the role of generating energy. For example, if the gravitational constant and the fine structure constant are smaller than their usual values, black holes can provide viable energy sources. In fact, all universes can support the existence of stars, provided that the definition of a star is interpreted broadly. For example, degenerate stellar objects, such as white dwarfs and neutron stars, are supported by degeneracy pressure, which requires only that quantum mechanics is operational. Although such stars do not experience thermonuclear fusion, they often have energy sources, including dark matter capture and annihilation, residual cooling, pycnonuclear reactions, and proton decay [24].

There's another important question about those fundamental "constants", like the speed of light and the gravitational constant. Are they really constant—even in our own universe? There's no obvious reason why they should be—and some people have even come up with reasons why they shouldn't be.

One such was the physicist Paul Dirac, who—as described in "The Relativity of Wrong"—was name-dropped in several SF stories by James Blish. In 1937, Dirac made the following point in a letter to the journal *Nature*:

The ratio of the electric to the gravitational force between electron and proton, which is about 10^{39}, and the ratio of the mass of the universe to the mass of the proton, which is about 10^{78}, are so enormous as to make one think that some entirely different type of explanation is needed for them... This suggests that the above-mentioned large numbers are to be regarded, not as constants, but as simple functions of our present epoch [25].

In other words, Dirac was suggesting that the so-called constants of the universe are actually variables that change their values on cosmological timescales. This idea was picked up by several other theoreticians, notably Carl Brans and Robert Dicke of Princeton University. They came up with an alternative to Einstein's theory of General Relativity in which the strength of gravitation varies over time. Quoting from a textbook by Steven Weinberg:

> Dirac's theory inspired a number of attempts to formulate a field theory of gravitation in which the effective "constant" of gravitation is some function of a scalar field… The most interesting and complete scalar-tensor theory of gravitation is that proposed by Brans and Dicke in 1961… In this theory, the gravitational constant G is replaced with the reciprocal of a scalar field [26].

As a theoretical idea this is all very well—but is there any observational evidence for time variation? A longstanding proponent of the notion of "inconstant constants" is the British physicist John Barrow. Writing in *Scientific American* in 2006, he explained how he and his team had looked for variations in the "fine-structure constant", a dimensionless number that describes the strength of the electromagnetic force:

> We anticipated establishing that the value of the fine-structure constant long ago was the same as it is today; our contribution would simply be higher precision. To our surprise, the first results, in 1999, showed small but statistically significant differences. Further data confirmed this finding. Based on a total of 128 quasar absorption lines, we found an average increase of close to six parts in a million over the past 6 to 12 billion years [27].

It has to be said, however, that other researchers have tried and failed to replicate Barrow's results. In any case, that variation of six parts in a million— over pretty much the entire life of the universe—is hardly a dramatic one.

Attempts to detect long-term variations in the gravitational constant have also drawn a blank. One team from the California Institute of Technology and the University of California did find an unusual effect, however, as they reported in a paper titled "Measurements of Newton's Gravitational Constant and the Length of Day", which appeared in *Europhysics Letters* in 2015:

> About a dozen measurements of Newton's gravitational constant, G, since 1962 have yielded values that differ by far more than their reported random plus systematic errors. We find that these values for G are oscillatory in nature, with a period of $P = 5.899 \pm 0.062$ years… However, we do not suggest that G is actually varying by this much, this quickly, but instead that something in the

measurement process varies. Of other recently reported results, to the best of our knowledge, the only measurement with the same period and phase is the length of day [28].

Many people may be surprised to hear that the "length of day" varies—after all, the Earth's diurnal rotation has formed the basis of precision time-keeping for centuries. Nevertheless, there are tiny variations in the rate of this rotation that are attributable to various geophysical processes.

As it happens, this very subject was discussed in a factual article in *Astounding Science Fiction* in November 1949. This was the work of astronomer R. S. Richardson, of the Mount Wilson and Palomar Observatories, who wrote:

> During the latter half of the 19th century the Earth ran fast, apparently gaining at the rate of a second a year until by 1897 the rest of the universe seemed to have dropped behind about 30 seconds… Then something happened to reverse the trend so that the Earth began to lose time. But around 1917 something happened to cause the Earth to start gaining again [29].

This was just one of dozens of science-fact articles that Richardson wrote for *Astounding* throughout the 1940s and 1950s. During the same period, he also produced a similar amount of science fiction under the pen-name of Philip Latham. One of the best known of the Latham stories, "The Xi Effect", appeared in *Astounding* in January 1950—just two months after that "length of day" article.

By coincidence (though of course we wouldn't have mentioned it otherwise), "The Xi Effect" deals with the topic of different physics. In the story, this is wrapped up in a pseudoscientific concept called "Xi space", attributed to a fictional character named Karl Gustav Friedmann (an odd choice of name, given the existence of the famous real-world astrophysicist Alexander Friedmann). Quoting from the story:

> Friedmann considered the familiar everyday world to constitute merely a tiny corner or "clot" in a vastly higher order of space-time or "Xi space". Ordinarily, events in the Xi space are on too gross a scale to exert a sensible effect on the fine-grained clot space. On rare occasions, however, a clot might be seriously disturbed by events of an exceptional nature in the Xi space, in somewhat the same way that the atoms on the surface of a stick of amber may be disturbed by rubbing it vigorously. When events in the super-cosmos happen to intrude upon an individual clot extraordinary results ensue; for example, angular

momentum is not strictly conserved, and Hamilton's equations require modification, to mention only a few.

It's amusing that—either by accident or design—the two examples cited, involving angular momentum and Hamilton's equations, are unlikely to mean anything to someone who isn't a professional physicist or engineer. In a similar vein, the story includes a footnote with a formal reference to *"Journal of the Optical Society of America*, vol. 30, p. 225, 1930".

Returning to the storyline—it soon becomes obvious that in this particular case, a brush with Xi space is causing the laws of physics to change in a radical way:

> Arnold reached for the pencil and a pad of yellow scratch paper. "Assume that this line represents the boundary of our local universe or clot," he said, drawing an irregular closed figure with a dot near the centre. "According to Friedmann, occasionally some disturbance in the outer super-cosmos or Xi space becomes sufficiently violent to affect a particular clot. Now there are several things that can happen as a result, but by far the most probable is that the clot will begin to shrink, very slowly at first and then more rapidly. But for a long time nobody would be aware of the shrinkage because everything within the clot shrinks in proportion, with one exception. That exception is the wavelength of electromagnetic radiation…"
>
> "I think I'm beginning to get it," said Stoddard, studying the diagram. "We didn't get any transmission beyond 20,000 angstroms because there wasn't any radiation to transmit."
>
> "That's it ! Our universe only had a diameter of twenty thousand angstroms. All radiation of longer wavelength was cut out."
>
> "About one ten-thousandth of an inch," said Stoddard, doing some fast mental arithmetic. He chuckled. "No wonder old Fosberg was worried!"
>
> "You see the Xi effect does give a consistent explanation of all the phenomena," said Arnold triumphantly. "In any case, we can't be in doubt much longer."
>
> "How's that?"
>
> "Why because the universe will have shrunk so much the optical spectrum will be affected. The landscape will change colour." [30]

That is indeed what happens—and to all intents and purposes, it's the end of the world. As a matter of editorial policy, *Astounding* rarely printed stories with downbeat endings—but this is one of the few exceptions.

As an antidote to such gloominess, let's move quickly on to everyone's favourite brand of "different physics": the kind that prevails in the world of animated cartoons. On the face of it, anything is possible in a cartoon—but

Fig. 4 For many people, the laws of "cartoon physics" are easier to understand than those of the real world (Wikimedia user Greg Williams, CC-BY-SA-2.5)

certain conventions are generally adhered to. Most famous of all is the cartoon law of gravity, which states that a character stepping off a cliff only starts to fall *after* they become aware of their situation (see Fig. 4).

The Spectrum of Fake Physics

In the course of this book, we've encountered a whole range of topics that could be classed, in one way or another, as "fake physics". The first chapter considered "Science Fiction Posing as Science Fact"—a relatively rare sub-genre of SF, of which Asimov's "Endochronic Properties of Resublimated Thiotimoline" is the best known example. This had all the outward appearance of a turgidly written scientific paper—albeit one that happened to be published in a magazine devoted to science fiction.

The second chapter described Asimov's notion of the "The Relativity of Wrong"—the idea that scientific hypotheses gradually creep closer and closer to a true picture of the way the world works. In this spirit, the field is wide open to speculative hypotheses—either in real science or SF—that *might* be true, although there is no real evidence for them. Such hypotheses, including things like faster-than-light tachyons and traversable wormholes, constitute another variety of fake physics.

Speculative hypotheses, however, do at least hang together logically. The same can't be said for the fake physics covered in the third chapter, "The Art of Technobabble". This showed how the language of physics, from jargon and mathematical equations to the highly formulaic style of the academic paper, can easily be simulated in a convincing if ultimately nonsensical way.

The fourth chapter looked at "Spoofs in Scientific Journals". While many of these are similar in style and purpose to Asimov's thiotimoline paper, there is one key difference. Rather than an SF magazine, they originally appeared in publications (some serious, others not-so-serious) aimed at an audience of professional scientists.

As in all other fields, the appearance of physics-related spoofs peaks dramatically around the 1st of April each year. Accordingly, the fifth chapter was devoted to a detailed examination of the "April Fool" phenomenon. As far as physics is concerned, the natural home of April Fool spoofs these days can be found in the arXiv repository of online "e-prints".

All the spoofs discussed up to this point were designed purely for entertainment. The ones in the sixth chapter, "Making a Point", are equally entertaining—but their authors had another, ulterior motive for creating them. This category of fake physics includes the famous Sokal hoax, designed to catch out editors working to a political agenda, and numerous sting operations against the new breed of highly profitable but poor quality "predatory" journals.

Finally, the present chapter has looked at examples of physicists—and scientifically literate SF authors—"Thinking Outside the Box". Although the physics here has undeniably "fake" aspects, it does carry a serious scientific message—whether it is the scientific debunking of paranormal claims, or thought experiments that would be completely impractical in the real world.

A selection of specific examples taken from throughout the book is shown in Fig. 5. For clarity, these have been arranged as a kind of two dimensional map (although there in no rigorously objective logic behind the placement of items).

The horizontal axis refers to the medium in which the item appeared, with serious scientific journals on the left and pure science fiction on the right. Several items lie between the two extremes—for example SF that contains a significant amount of scientific exposition (such as Carl Sagan's *Contact*, or Stanisław Lem's "The New Cosmogony") or "factual" pieces that touch on science-fictional themes (all the way from Alcubierre's warp drive to the CERN press release about the Force).

The vertical axis relates to the authors' intention in creating the piece, ranging from just-for-fun at the top—things like the turboencabulator or Asimov's thiotimoline—to a more serious purpose at the bottom. In the latter case, a further distinction is necessary. Some pieces that were created with a serious purpose—such as R. V. Jones's wartime disinformation, or Michael Crichton's climate-change-denying *State of Fear*—are nonetheless based on bogus science. Others, such as the winners of the Ig Nobel prize, or the George Gamow's Mr Tompkins stories, do convey a serious scientific message despite

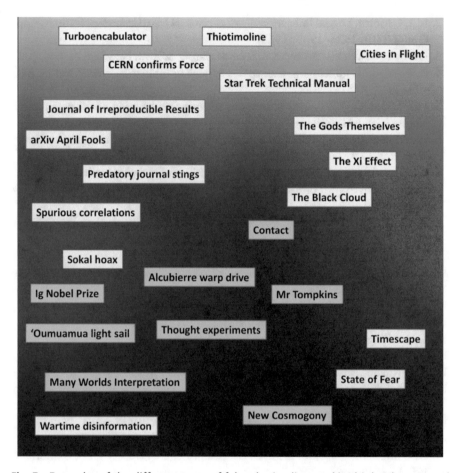

Fig. 5 Examples of the different types of fake physics discussed in this book, arranged as a two-dimensional "map". The horizontal dimension extends from serious scientific journals on the left to pure science fiction on the right, while the vertical dimension runs from "just for fun" items at the top to those with more serious intent at the bottom. Items highlighted in green, while having some "fake" attributes, do nevertheless contain some valid science

their whimsical nature. For clarity, therefore, the latter have been highlighted in green.

References

1. M. Planck, Natural measurement units. Proc. R Prussian Acad. Sci. **5**, 479–480 (1899)

2. M.J. Carlotto, Digital imagery analysis of unusual Martian surface features. Appl. Opt. **27**(10), 1926–1933 (1988)

3. F. Hoyle, C. Wickramasinghe, *Lifecloud* (Harper & Row, New York, 1978), p. 13

4. P. Weber, J.M. Greenberg, Can spores survive in interstellar space? Nature **316**, 403–407 (1985)

5. J.G. Taylor, E. Balanovski, Is there any scientific explanation of the paranormal? Nature **279**, 631–633 (1979)

6. F.J. Tipler, Rotating cylinders and the possibility of global causality violation. Phys. Rev. D **9**, 2203–2206 (1974)

7. Entry for "Rotating cylinders and the possibility of global causality violation" in Larry Niven's bibliography. http://news.larryniven.net/biblio/display. asp?key=124

8. B. Clegg, *How to Build a Time Machine* (St Martin's Press, New York, 2011), pp. 183–184

9. J. Al-Khalili, *Paradox* (Black Swan, London, 2013), p. 196

10. I. Asimov, *Biographical Encyclopedia of Science and Technology* (Pan Books, London, 1975), p. 399

11. J. Al-Khalili, *Paradox* (Black Swan, London, 2013), p. 106

12. J. Al-Khalili, *Paradox* (Black Swan, London, 2013), p. 126

13. S. Hawking, *Black Holes* (Bantam Books, London, 2016), p. 7

14. S. Hawking, *A Brief History of Time* (Bantam, London, 1988), pp. 87–88

15. J. Al-Khalili, *Paradox* (Black Swan, London, 2013), p. 154

16. Michael Wilson & Rod Serling (screenplay), Planet of the Apes (20th Century Fox, 1968)

17. P. Boulle, *Planet of the Apes* (Vintage Books, London, 2011), p. 12

18. L. Ron Hubbard, *Return to Tomorrow* (Ace, New York, 1954), p. 70

19. G. Gamow, *Mr Tompkins in Paperback* (Cambridge University Press, Cambridge, 1993), pp. xv–xvi

20. G. Gamow, *Mr Tompkins in Paperback* (Cambridge University Press, Cambridge, 1993), pp. 1–8

21. M. Kaku, *Parallel Worlds* (Penguin, London, 2005), pp. 247–249

22. I. Asimov, *The Gods Themselves* (St Albans, Panther, 1973), pp. 21–23

23. S. Baxter, Raft. Infinity Plus. http://www.infinityplus.co.uk/stories/raft.htm

24. F.C. Adams, Stars in other universes: stellar structure with different fundamental constants. J. Cosmol. Astropart. Phys. (August 2008). https://arxiv.org/abs/0807.3697

25. P.A.M. Dirac, The cosmological constants. Nature **139**, 323 (1937)

26. S. Weinberg, *Gravitation and Cosmology* (Wiley, New York, 1972), p. 622

27. J.D. Barrow, J.K. Webb, Inconstant constants. Scientific American, (February 2006). https://www.scientificamerican.com/article/inconstant-constants-2006-02/

28. J.D. Anderson et al., Measurements of Newton's gravitational constant and the length of day. Europhys. Lett. (April 2015). https://arxiv.org/abs/1504.06604

29. R.S. Richardson, The time of your life. Astounding Science Fiction, (November 1949), pp. 110–121
30. P. Latham, *The Xi effect, Best SF* (Faber, London, 1972), pp. 125–147

Appendix: Science for Crackpots

As a bonus here is a short spoof written a few years ago by the author of this book. It originally appeared online on the *Mad Scientist Journal* website in October 2014, and was reprinted in *Mad Scientist Journal anthology #11* (DefCon One Publishing, Kindle edition, 2014).

Science for Crackpots?

I know what you're thinking: "Crackpots don't need to read textbooks because they already know everything". You're right, of course. But this isn't a textbook, just a collection of useful tips based on my own thirty years of experience as one of the world's leading crackpots.

The Scientific Method

The heart of mainstream science is something they call "the scientific method". This is a kind of Masonic handshake that scientists use to keep insiders inside and outsiders outside. In reality the scientific method is like the Emperor's New Clothes—it sounds fancy but there's nothing really there. As a crackpot, you won't lose any credibility if you ignore the scientific method altogether.

Many promising young crackpots are put off by the mistaken belief they have to plough through piles of books with boring titles like "Integrated

© Springer Nature Switzerland AG 2019
A. May, *Fake Physics: Spoofs, Hoaxes and Fictitious Science*, Science and Fiction,
https://doi.org/10.1007/978-3-030-13314-6

Principles of Zoology" or "Principles and Applications of the General Theory of Relativity". This simply isn't the case. These books were written by scientists, and scientists don't know everything. If they did, the world wouldn't need crackpots.

One of the founders of quantum theory, Max Planck, observed that "new scientific ideas never spring from a communal body, however organized, but rather from the head of an individually inspired researcher." So where will the next great scientific idea come from? It might be your head—in fact, it probably will be.

Belief Systems

One of the first things the aspiring crackpot needs to do is choose a belief system. Be as imaginative as you want—you can always change it later if you decide you don't like it. As the great philosopher Rudolf Steiner pointed out, "Truth is a free creation of the human spirit, that never would exist at all if we did not generate it ourselves."

The belief system of modern science has passed its use-by date. Many of its ideas are still rooted in the avant-garde aesthetics of the early twentieth century. A perfect example is Einstein's Theory of Relativity, which is a slap in the face to common sense in the same way that Picasso's cubist style—which emerged at the same time—was a slap in the face to representational art. Just as the art establishment deified Picasso, the science establishment deified Einstein.

Fortunately the world of ideas is a free market, and you can pick and choose what you wish to believe or disbelieve. Most crackpots feel an instinctive aversion to Einstein's Theory of Relativity, and can easily come up with a simpler and more satisfying theory of their own. In fact, there is a standard Relativity Test that can be applied to aspiring crackpots: If an individual feels no urge whatsoever to come up with an alternative to the Theory of Relativity, they will probably never make a good crackpot.

Scientific Dogma

Mainstream scientists take a relentlessly dogmatic approach to methodology. They insist on starting with the evidence, and then looking for a theory to explain that evidence. With a mindset like that, it's no wonder science makes such slow progress.

On the other hand, crackpots are more open minded. They avoid the dogma trap by starting with an explanation and then looking for evidence to support it. That's why the field of ufology has made such enormous progress in the last sixty years. If you adopted the mainstream approach, it would mean examining all the eyewitness testimony and leaked government documents and YouTube videos, then formulating an exhaustive set of mutually exclusive hypotheses and going through the laborious process of testing them in a scientifically rigorous way. You wouldn't get anywhere at all—which is why mainstream science hasn't made any progress whatsoever in understanding the UFO phenomenon.

On the other hand, if you take the more rational, level-headed approach of assuming from the start that UFOs are interstellar spacecraft piloted by small grey humanoids from Zeta Reticuli, and then fitting all for available evidence to the theory, you will know exactly what we know today: that UFOs are interstellar spacecraft piloted by small grey humanoids from Zeta Reticuli.

Falsifiability

Of all the concepts in the philosophy of science, perhaps the most bizarre is that of the falsifiable hypothesis. "Falsifiable" sounds like a bad thing, but—in the eyes of mainstream science—it's actually a good thing. Why on earth is that? What person in their right mind would want their pet hypothesis to be falsifiable? Yet that's exactly what scientists want – most of them won't even look at non-falsifiable hypotheses. They refer to them as "Not Even Wrong", and make it sound like a pejorative. But right and wrong are opposite extremes, so "Not Even Wrong" is somewhere in the grey area between the two. That's the best place for a crackpot to be.

Wherever possible, crackpots should stick to non-falsifiable hypotheses and leave the defeatist, angst-ridden concept of falsifiability to the mainstreamers. The latter may formulate a hypothesis such as "Gigantopithecus is extinct", since it is falsifiable. If a Gigantopithecus is caught on a YouTube video, that proves it's not extinct and so the hypothesis is false. On the other hand, a crackpot is better off with a non-falsifiable hypothesis such as "Gigantopithecus is not extinct". No matter how much you search the woods without finding Bigfoot, that doesn't disprove your hypothesis. "Absence of Evidence is not Evidence of Absence," as Galileo said to the Spanish Inquisition.

A Role Model

Ah, yes—Galileo. He's my role model, and I bet he's your role model too. Galileo Galilei (1564–1642) lived at a time when, like today, the establishment authorities were blinkered, short-sighted fools. They believed the Sun orbited around the Earth, and they believed heavy objects fell faster than light objects. They were wrong, and Galileo told them they were wrong, but they stuck to their erroneous views and persecuted him into the bargain. That's a situation most of us can relate to – but we can take heart in the fact that history has vindicated Galileo and made a laughing-stock of his persecutors.

Just like you, Galileo was told his ideas were nonsense. And just like Galileo, you will be proved right in the end.

Reporting Your Results

Having come up with your revolutionary new theory, you will of course want to inform the world about it. There's a right way and a wrong way to go about this. The wrong way is to post a Facebook status along the lines of: "Hey, guess what you guys!!! You CAN travel faster than light!!! I just PROVED it!!! Albert Einstein was a LOSER!!!"

Strange as it may seem, the mainstream scientific community usually ignores devastatingly important announcements of this type. If you want them to take your work seriously, you have to write it up in the correct academic style—with plenty of footnotes, references, equations and diagrams. This may sound daunting, but it's not that difficult when you get the hang of it. To mainstream scientists, style is more important than substance. Ultimately, the whole issue of scientific credibility comes down to the effective use of jargon.

For the crackpot, one of the most insidious aspects of mainstream science is the concept of peer review. It wouldn't be a problem if the reviewer really was a peer, i.e. another crackpot. In practice, however, it's more likely to be a member of the scientific Thought Police with a vested interest in suppressing the work of outsiders. For this reason, most crackpots choose to boycott the peer review process altogether.

Occam's Razor

William of Occam, who lived around 700 years ago, made a famous observation to the effect that "it is futile to do with more things that which can be done with fewer". This gave rise to the principle of Occam's Razor, which is one of the fundamental tenets of the scientific method. In simple terms, it says that if there are two competing hypotheses which describe the facts equally well, then the simpler theory is more likely to be correct.

It's obvious that by espousing Occam's Razor, scientists are shooting themselves in the foot. Mainstream theories are always obscure and overcomplicated, while crackpot ones are simple and straightforward. So crackpots will win out on the grounds of Occam's Razor every time.

As an example, consider the field of ufology. Scientists are always falling over themselves to debunk UFO sightings, but they do it "with more things rather than fewer". As such they are automatically wrong, by a simple application of Occam's Razor. They invoke swamp gas, weather balloons, the planet Venus, flocks of pelicans, attention-seeking hoaxers and mentally retarded witnesses. That's six different explanations, and they don't stop there. On the other hand, ufologists only need one explanation, i.e. that a UFO is exactly that—an unidentified flying object, namely a piloted space vehicle from another planet.

Paradigm Shifts

The concept of a paradigm shift, like Occam's Razor, is one of the basic elements of the scientific method. But again, by drawing attention to the subject scientists are effectively shooting themselves in the foot. Shifting paradigms is something crackpots are much better at doing than their mainstream counterparts.

A paradigm is the overarching framework of a scientific theory, the shifting of which is the whole point of crackpot science. Even a fairly average practitioner may be capable of shifting one or two paradigms per year. I once shifted three paradigms in a single week. In contrast, many establishment scientists will go through their whole career without shifting even one paradigm.

Quantum Theory

Technically Quantum Theory is a branch of physics, but it's quite unlike any of the others. It doesn't involve any authoritarian "laws of physics" that you're not allowed to break. Relativity says you can't travel faster than light. The Second Law of Thermodynamics says you can't have perpetual motion. Quantum Theory, on the other hand, says you can do anything you like.

To add to the good news, Quantum Theory doesn't involve any mathematical equations at all. It's based entirely on jargon, which you can use to mean whatever you want it to mean. A few examples of quantum jargon are: "Nonlocality", "Entanglement", "Wave-Particle Duality", "Hidden Variables" and "the Uncertainty Principle". Feel free to use these terms in any way you want: no-one else understands them any better than you do.

Although Quantum Theory was developed by mainstream scientists, they don't understand it half as well as crackpots do. They haven't got a clue how to apply it to the real world, either. In the hundred years that quantum theory has existed, mainstream science hasn't come up with a single practical application of it. In the meantime, crackpots have come up with dozens of quantum applications, mainly in the fields of mysticism and alternative medicine. Much of this is due to the work of people like Deepak Chopra, who discovered the link between quantum theory and the ancient Indian tradition of Ayurveda, and his identical twin Fritjof Capra, who discovered the link between quantum theory and the ancient Chinese tradition of Taoism.

Relativity

"No, not relativity!" I hear you cry. "Relativity is a preposterous concept with no place in a serious work of crackpot science."

Well, yes and no. There's relativity and there's relativity. It's all relative, you see.

Einstein's Theory of Relativity, based on the ridiculous notion that the laws of physics are the same in all inertial frames, is obviously wrong and no more needs to be said about it. Many crackpots break into a homicidal rage at the very mention of Einstein's name, and it's easy to see why.

But there is another kind of relativity—the relativity of ideas. This is a different matter altogether. Despite the best efforts of mainstream scientists to suppress the fact, there's no denying it: All ideas are equally valid.

Scientific Objectivity

A characteristic feature of the scientific method is its relentless insistence on objectivity. If something can't be put under a microscope in a laboratory, then as far as science is concerned it doesn't exist. This is like throwing the baby out with the bathwater. It automatically ignores the whole field of anomalous phenomena—which, by definition, can't be pinned down by objective analysis. Subjects like ufology, cryptozoology and the paranormal are characterized not by cold laboratory data but by anecdotal evidence – something traditional science simply isn't equipped to handle.

Mainstream scientists have a supercilious contempt for eyewitness reports, no matter how numerous or persuasive they are. For example, there is overwhelming evidence that an alien spacecraft crashed at Roswell in 1947, based on the testimony of countless eyewitnesses—some of whom weren't even born at the time. Yet science doggedly turns a blind eye to such evidence. The same is true of more or less any Bigfoot encounter you care to name.

Crackpots should not, of course, make the same mistake as the mainstreamers. Always listen to the witness. Doubting someone's testimony is tantamount to calling them a liar, which could leave you open to libel action. The safest thing is to believe everything you're told.

Unsolved Problems

Science is full of unsolved problems. Here are some of them:

- How did the heads of arthropods evolve into their present form?
- Why do molecules exhibit increased nucleophilicity in the alpha effect?
- Under what conditions do solutions of the Navier-Stokes equations exist?
- Why is the cosmic microwave background aligned with the ecliptic?
- Does the Schrödinger equation have an exact solution for Helium?

These questions are all rubbish. Who cares about any of them? It's because scientists waste their time on this sort of thing that the world needs crackpots to address the really important problems. Things like this, for example:

- What star system did the UFO that crashed at Roswell come from?
- Is the Yeti the same species as Bigfoot or a different one?
- Does the effectiveness of telepathy depend on the distance involved?

- Why does the temperature drop when a ghost is in the room?
- How many members of the US Congress are shape-shifting reptiles?

If mainstream science addressed questions like these, people might start taking it seriously.

Concluding Remarks

It's a sad fact that crackpots are a persecuted minority. You may not be persecuted to the same extent that Galileo was, but you're bound to be persecuted at some level. People on internet forums and social media will tell you that you're wrong. They will say you're crazy and incompetent and ignorant and deluded. They will try to suppress your work, and if they can't do that they will resort to making fun of it. These people are Government agents, and they use the same tactics on anyone who gets too close to the truth. You just need to develop a thick skin like any other crackpot.

If all else fails, remember Galileo. They said Galileo was mad, but he was proved right in the end. If you get called mad often enough you will be proved right too, just like Galileo. If people don't call you mad, then you're not trying hard enough. You should consider reformulating your theory in UPPERCASE CHARACTERS so the world will finally understand its earth-shattering significance.

Printed in the United States
By Bookmasters